ARCHITECTS OF RUIN

ALSO BY PETER SCHWEIZER

Makers and Takers

Do As I Say (Not As I Do): Profiles in Liberal Hipocrisy

The Bushes: Portrait of a Dynasty

Reagan's War: The Epic Story of His Forty-Year
Struggle and Final Triumph over Communism

Chain of Command

Landmark Speeches of the American Conservative Movement

Friendly Spies: How America's Allies Are Using
Economic Espionage to Steal Our Secrets

Fall of the Wall: Reassessing the Causes and
Consequences of the Collapse of the Berlin Wall

Victory: The Reagan Administration's Secret Strategy
That Hastened the Collapse of the Soviet Union

ARCHITECTS OF RUIN

How Big Government Liberals Wrecked
the Global Economy—and How They Will
Do It Again If No One Stops Them

PETER SCHWEIZER

HARPER

NEW YORK · LONDON · TORONTO · SYDNEY

HARPER

A hardcover edition of this book was published in 2009 by HarperCollins Publishers.

HarperCollins books may be purchased for educational, business, or sales promotional use. For information please write: Special Markets Department, HarperCollins Publishers, 10 East 53rd Street, New York, NY 10022.

FIRST HARPER PAPERBACK PUBLISHED 2010.

Library of Congress Cataloging-in-Publication Data has been applied for.

ISBN 978-0-06-195337-8

10   11   12   13   14   DIX/RRD   10   9   8   7   6   5   4   3   2   1

For Rochelle

CONTENTS

INTRODUCTION
A Failure of Capitalism—or Liberal Social Engineering?

If you happened to be watching the MSNBC news channel in early January 2009, there's a good chance you might have fallen off the sofa after reading the crawl on the bottom of the screen: "ECONOMY PROJECTED TO SHRINK 202%."

What?

It was a misprint, of course. But to many frightened viewers this nightmare scenario might not have seemed far-fetched. As the wheels started to shake and come off the American economy in mid-2008—at the height of the presidential election—there was concern verging on panic about the possibility of a total economic meltdown. Thus the *Los Angeles Times* asked "Could Another Great Depression be Lurking on the Horizon?" The *Chicago Tribune* offered the banner "75 Years Later, a Nation in Crisis Again." *Time* magazine wondered: "The End of Prosperity?" And from the usually sober *Wall Street Journal*: "Worst Crisis Since '30s, with No End in Sight."

The concern and panic were real because the numbers were real.

Loan defaults and bank foreclosures had skyrocketed. Real estate values collapsed nationwide, trapping many people in homes they could not sell and struggling to pay mortgages. Trillions of dollars in value were wiped out of the stock market, mutual funds, and pension funds. Millions of Americans were suddenly unable to afford retirement, their savings wiped out and the value of their retirement accounts effectively cut in half. A systemic credit market freeze was temporarily thawed by the printing of massive amounts of money by the federal government. Large financial entities such as Bear Stearns and Lehman Brothers effectively collapsed. Others, such as Citigroup, had to be pumped full of taxpayer money to keep them functioning.

The contagion quickly spread to the international system. Tiny Iceland's economy completely imploded, and in central Europe and the developing world, markets tottered on the brink of collapse. Around the world a clamor arose for the Americans to save the global system.

Barack Obama came to office pledging to turn the economy around and "create or save" millions of jobs while at the same time enacting a "transformative" social agenda. A terrified Congress rammed through an enormous (and pork-laden) stimulus package, promising that it would stabilize the economy and restore hope to millions of out-of-work Americans. Yet six months later it seemed to be having little effect. Total job losses approached 6 million as unemployment edged toward double digits and welfare rolls expanded. The federal budget deficit crossed $1 trillion in 2009, and many state budgets were equally deep in the red. California was on the brink of bankruptcy. We now appear destined for massive federal tax hikes to pay for the bailouts and the underwriting of President Obama's ambitious social agenda. State-level tax hikes and cutbacks will undoubtedly follow.

And it's not over—not by a long shot. Experts foresee another wave of mortgage defaults and predict that unemployment will continue to rise. In Washington the answer to every problem seems to be spend more money. No one seems to know what else to do. Meanwhile, there is talk that the dollar may cease to be the international reserve currency. There are also calls abroad for tighter global regulation and the creation of transnational bodies that will direct the international economy—a fatal blow to the economic supremacy that Americans have taken for granted since World War II.

All of this has led Americans to wonder: What happened? How the heck did we get here? Whose fault is it? Who do we blame? What mistakes were made? How can we get out of this mess?

There has been much debate about this question, but the ultimate source of the problem, it is generally agreed—the triggering event that caused the chain of other dominoes to fall—was the collapse of the subprime mortgage market in the United States. Banks and mortgage companies had made trillions of dollars in loans to individuals with terrible credit. They signed loans with illegal immigrants, offered so-called NINJA (No Income, No Job, No Assets) mortgages, and allowed people with bad credit to leverage their money. When the loans began to fail in large numbers, a new term entered our national vocabulary: toxic assets. And so the crisis began.

Still, an underlying mystery remained: What explains this perplexing behavior? Were they nuts? Did they simply take leave of their senses?

The conventional narrative was written in the first days of the collapse. And as usual, the loudest, most obstreperous voices seemed to prevail. "The private sector got us into this mess," Congressman Barney Frank indignantly declared as events began to unfold; "the government has to get us out of it." [1]

According to this view, deregulation of the banking industry had encouraged the rise of "predatory lenders" who had pushed home loans on people who couldn't afford them. Those loans were then sold to unscrupulous Wall Street financiers, who repackaged them in the form of mortgage-backed securities. The securities were sold in turn to mutual funds, pension funds, and various foreign investors. But their value was grossly overstated and ultimately rested on the faulty assumption that housing prices would keep rising indefinitely. Once again, the supposed result of irresponsible deregulation of financial markets.

This explanation, coming from Frank, had the obvious benefit of pinning the collapse on his political enemies, the Republicans, while completely exonerating any Democrat (such as himself) who had responsibility for overseeing Fannie Mae and Freddie Mac, the government-backed lending institutions that traditionally acted as a backstop to the housing market. It is not an accident that Frank has been in the forefront of attempts to minimize the crisis or (when it could no longer be denied) deflect the blame to his opponents. When some conservatives pointed out that Fannie and Freddie had abandoned their sober mission of stabilizing the middle-class housing market in favor of a misguided crusade to expand minority home ownership by forcing banks to lower their lending standards, Frank and his allies brazenly shouted them down.

Meanwhile, a chorus of authoritative voices in the press rushed to second the indictment. Nouriel Roubini of New York University blamed "unregulated free market zealotry" for the economic collapse. The Nobel Prize-winning economist Paul Krugman, who also writes a column for the *New York Times*, denounced "the great unraveling" of economic regulations pushed by conservatives. Another Nobel laureate, Paul Samuelson, claimed that the financial meltdown was an inevitable consequence of free-market economics. "I wish Milton Friedman were still alive," he said, "so he could witness how his extremism led to the defeat of his own ideas."[2] The economist Jeff Madrick of the New School for Social Research in New York City likewise declared that the crisis represented "the end of the age of Milton Friedman."

Many on the left started using a new term: "free-market fundamentalists." First coined by the billionaire currency speculator George Soros, today the term is applied to anyone who disagrees with the conventional narrative that unregulated capitalism is inherently destructive. Thus Harold Meyerson explained in the *Washington Post* that "market fundamentalism" had "totally failed" and suggested that we look instead to the German model for a "form of capitalism [which] has proved more sustainable than Wall Street's."

Democrats immediately saw the potential of the crisis to tip the scales in favor of more government intervention in the economy. Lawrence Summers, a former Treasury secretary under Bill Clinton who was now Obama's chief economic adviser, predicted that now "the pendulum will swing—should swing—towards an enhanced role for government in saving the market system from its excesses and inadequacies."[3] White House Chief of Staff Rahm Emanuel declared that it would be a shame to let the crisis go to waste.

Emanuel had good reason to anticipate success. Coming at the height of the 2008 election, the Wall Street meltdown tipped the balance heavily in favor of Barack Obama, creating a Democrat majority in both houses of Congress and effectively giving the new president a blank check—in his view—to rewrite the American social contract.

As a result, the rush to heavy government intervention, new programs, and massive spending was now treated as inevitable. It was the 1930s all over again, and Obama was the new FDR. Free-market economics had been tried and found wanting. Obama referred to its theories dismissively as "failed ideas" and refused to entertain any talk of tax cuts or (God forbid) "doing nothing" in response. To the contrary, the crisis proved that it was time to return to stronger government controls. Anyone standing in the way was seen as part of the political fringe, a die-hard ideologue on par with a Holocaust denier.

This is the self-serving fairy tale propounded by Barack Obama and his allies in Congress and the press. The actual truth about what happened was a much more interesting and complicated—and incriminating—story, too complex to be conveyed in a media sound bite. So it is not surprising that this simplistic story line has become the accepted wisdom.

Yet the chorus of liberal triumphalism that sought to usher in a "new" New Deal ignored some inconvenient facts.

1. *The American capitalist system is not unregulated.* Given what you have heard in the press, it may surprise you to learn that the banking sector is in fact the most regulated industry in the United States. The nuclear industry is a possible exception.

2. *America has not experienced massive deregulation over the past decade.* Strange that his liberal critics should speak of President George W. Bush as a free-market radical, "slashing" and "gutting" regulations. In fact, the Bush administration saw new regulations added to the Federal Register at the rate of a thousand pages a year. If this is deregulation, what would regulation look like?[4] (Full disclosure: I served as a consultant to the White House Office of Presidential Speechwriting in 2008–2009.)

3. *The federal government did not fail to "police" capitalism.* In fact, the opposite is true. The real culprit is not the market but the government; or, more precisely, a *socialist distortion* of the free market created by the government. Under pressure from liberal activists, the federal government created a series of "protected" markets that were insulated from external economic pressures in order to achieve political ends that could not be realized by other means.

Nor did this happen overnight. It was a massive social engineering project, a grand generational enterprise, thirty years in the making, carried out by an ad hoc alliance of radical activists, labor unions, liberal politicians, federal bureaucrats, and Wall Street financial titans who sought to make getting a mortgage and owning a home a civil right.

This is a story that has never been told, mainly because the actual extent of this ambitious multigenerational project has not been fully grasped. But it is vital that we learn it and take its lessons to heart, not just to understand the present crisis but to prevent its ever happening again.

The cast of characters includes three distinct but overlapping groups of people.

First came the community activists—idealists such as the young Barack Obama and national organizations such as ACORN—who, beginning in the 1960s, waged war against traditional lending practices in the name of fighting racism and poverty. Inspired by the legendary community organizer Saul Alinsky, their tactics were extremely confrontational, and their goals were frankly socialist: to take money from the rich (in this case banks and other lenders) and give it to the poor on very favorable terms, a process that they called "the democratization of capital" but that had more in common with a Mafia protection racket. I refer to this group as the Robin Hood Gang.

Still, the fair housing movement would probably have remained a fringe phenomenon had it not been for strong support from liberal politicians in Washington. Beginning in the 1970s, liberal lions such as Senators Ted Kennedy and William Proxmire oversaw the passage of legislation that empowered local activists to force banks into lowering lending standards. The Congressional Black and Hispanic Caucuses played their part by vociferously demanding new subsidies for poor and minority constituents. Finally, a phalanx of powerful liberals, including Bill Clinton, Robert Rubin, Andrew Cuomo, Janet Reno, Deval Patrick, Henry Cisneros, Barack Obama, Nancy Pelosi, Charles Schumer, Jon Corzine, and many others less well known, began pushing legislation to encourage or compel banks to make risky loans to individuals who should not have received them.

These people were part of the new wave of liberal activists turned political careerists who rose to power with the Clintons. These officials conspired to put the full power of the federal government behind the new civil rights crusade to expand minority home ownership— and often reaped political and financial rewards in the process. (Their resulting ethical problems and conflicts of interest nearly crippled the Obama administration at the outset.) As a result, billions of dollars' worth of loans were made to poor and minority applicants who would never have qualified under traditional lending criteria.

Meanwhile, the liberal baby boomers who had risen to the top on Wall Street pumped hundreds of billions of dollars into their efforts through the sale of so-called mortgage-backed securities and derivatives. These bankers, like their cohorts in Washington, believed they could do good *and* do well by extending credit to aspiring minority home owners via new forms of "creative" finance while at the same time racking up enormous fees and bonuses.

This three-headed movement of activists, liberal politicians, and socially conscious "do-good" capitalists on Wall Street got their wish. During the housing bubble, minorities accounted for twice as many subprime dollars borrowed per capita as did whites. Not surprisingly, the Federal Reserve of Boston found that they defaulted on subprime loans at twice the rate of whites.

To be fair, there were other important factors. Many have correctly pointed to the availability of cheap money, brought about by the loose monetary policies of the Federal Reserve under Chairman Alan Greenspan, as a major culprit. As Professor John Taylor of Stanford and others have argued, the growth in the money supply created cheap money and helped to inflate the real estate bubble. Greenspan has also been faulted for his policy of letting bubbles inflate and then repairing the damage rather than trying to prevent them. And when the Fed printed more money to increase liquidity as the crisis began, it made things even worse.[5]

But the heart of the story is the role that radical activists and liberal politicians in Washington played in trying to harness the U.S. financial system to advance their socialist agenda. Properly understood, it is a cautionary tale about the perils of trying to use the power of the state to do good, to help people by giving them a leg up, to "level the playing field." Ironically, such efforts have usually ended up doing the most harm to the very people they were intended to help. The result in this case was no different.

This book will illuminate the hubris and impatience of liberal baby boomers who for the most part have spent their careers in government, high finance, and social activism and as a result know almost nothing about how wealth is actually created in the system that has been entrusted to their care. Unlike earlier generations, who valued the free-market system (despite its flaws) for its ability to generate wealth and maximize liberty, the liberal baby boomers—born to affluence, burdened by guilt—saw the capitalist system as inherently flawed and unfair. More important, they saw it as a system that could, and should, be manipulated for "progressive" social purposes. Call it "do-good capitalism": the merging of sixties social values with the rewards of the profit system. The chief buzzwords of this enlightened form of capitalism are the fashionable notions of socially responsible investing and corporate citizenship.

As the liberal boomers rose to power in the 1970s, '80s, and '90s, they increasingly sought to harness the engine of capitalism to their vision of a good society. Thus, to further their activist goals, liberals in and out of Washington pushed the federal government deeper and deeper into engagement with the housing market, artificially driving the costs of lending down and pumping the system full of toxic debt.

At the same time, liberals in the Clinton administration entered into an unprecedented partnership with the financial industry that amounted to a form of state capitalism. Under Clinton, a series of Wall Street bailouts taught the big financial houses that if they failed, the federal government would come to their rescue. Almost all of the firms sucked up in the vortex of the financial crisis today were bailed out several times in the previous decade from wildly irresponsible investments. This only had the effect of further corrupting their judgment, inuring them to risks by insulating them from the ruthless discipline of the market.

Why has this story been missed? It isn't all that hard to understand.

First, of course, the operations of quasi-governmental agencies such as Fannie Mae and Freddie Mac are shrouded in layers of bureaucratic tedium. Few reporters take the time to actually dig into their inner workings. Besides, they are governed by congressional committees, so people naturally assume that someone responsible is paying attention. Few would have thought that the very politicians charged with overseeing the mortgage giants were themselves liberal activists busily engaged in stampeding them over a cliff.

Second, the housing activists, while certainly loud and annoying, have generally been perceived as a fringe phenomenon, or at best a local problem. How could a bunch of ragged-looking radicals shouting themselves hoarse in front of a bank or city agency present a threat to the global economy?

Finally, there was the big surprise on Wall Street. Most Americans could be forgiven for assuming that the heads of major Wall Street firms were thoroughly rational actors who ran their businesses with a cold eye to profit and the bottom line. The notion that they might have been corrupted by a new kind of government-sponsored capitalism that promised huge rewards for little risk would not have occurred to many people.

All of this is vital to understanding how we got into the current mess. It is extremely important that conservatives challenge the political fairy tale that the cause of the collapse was unregulated free-market capitalism. Such a partisan myth, once established in the media, is extremely difficult to dislodge.

But more important, this book is meant to be a cautionary tale—an urgent warning—about what the future will hold in an era of liberal economic dominance. The mortgage meltdown was but the first activist- and government-induced bubble; others will surely follow. We need to be constantly on guard against the liberals' urge to meddle in the economic system, using the power of the state to distort markets in order to advance their social goals.

That being the case, two additional facts should be very disturbing to American taxpayers.

First, the same people who caused the debacle have now been tasked with cleaning it up. The Obama administration is full of Clinton retreads, and they show no signs of having learned anything from the damage they have wrought. Do we really want them to be in charge of redesigning the economic system?

Second, the same cast of characters is busy leveraging state power to manipulate capitalism for their next great social cause: the so-called green economy. Just as occurred in the subprime mortgage crisis, federal authorities and environmental activists are working in tandem, browbeating energy companies and the automotive industry, using the power of the state to compel the creation of carbon-trading schemes and the forced development of green technologies that are simply not profitable. This approach essentially co-opts the regulatory power of the government to create false incentives to invest in green technologies.

The Silicon Valley investor Eric Janszen (who according to the *New York Times* accurately predicted the dot-com bubble) says that the hype and activism behind green technology will create enough "fictitious value" that the coming green tech bubble will reach an as-

tonishing $20 trillion . . . before it bursts. In the meantime, environ-
mental activists and their political allies stand to profit handsomely:
former Vice President Al Gore has already netted $100 million in
profits from green economy schemes.[6]

Plus, of course, the green agenda offers plenty of scope for good
old-fashioned political self-dealing. See, for example, this story from
the *Washington Times* of July 15, 2009: "Rep. Ed Perlmutter of Colorado inserted a provision into the recently passed House climate
change bill that would drum up business for 'green' banks, such as
the one he has invested in and his family and a political donor helped
found in San Francisco. . . . Mr. Perlmutter, a two-term Democrat,
has two investments in the 3-year-old New Resource Bank, which
calls itself the nation's first green bank."[7]

Needless to say, there will be much more to come. This is just the
tip of the iceberg.

We have not nearly seen the end of liberal activists trying to manipulate the capitalist system for their own profit and social goals.
Unless they are stopped, the rest of us are going to pay the price.

1

THE ROBIN HOOD AGENDA
How a Gang of Radical Activists and Liberal Politicians
Set the Stage for the Biggest Bank Heist in History

"We want it. They've got it. Let's go get it."

—GALE CINCOTTA

In 1994, the young Barack Obama, freshly out of Harvard Law School, joined two other attorneys in filing a lawsuit with the United States District Court for the Northern District of Illinois. The suit charged Citibank, the giant mortgage lender, with racism because it had denied home loans to several black applicants.

The details of *Selma S. Buycks-Roberson v. Citibank Federal Savings Bank* were pretty simple. Ms. Buycks-Roberson had applied for a loan but had been turned down "because of delinquent credit obligations and other adverse credit" (in the words of her legal team). She applied a second time but was again rejected. According to the lawyers, Ms. Buycks-Roberson was "incensed at the treatment" she received and "suffered great embarrassment, humiliation and emotional distress" at not getting the loan.[1]

Two more rejected applicants soon joined the suit. Like Ms. Buycks-Roberson, they contended that they had been turned down not for financial reasons but because they were black. Citibank "refused to approve Plaintiff's loan because Plaintiff is African-American," wrote her attorneys.[2] The lawsuit further claimed that "many dozens of African-American applicants" had likewise been rejected "because of race or color, or because of the racial composition of their neighborhood in which their property was located." Therefore, they declared, this would be a class-action lawsuit.

Claiming that Citibank had violated the Equal Credit Opportunity Act, the Fair Housing Act, and the Thirteenth Amendment (which had abolished slavery), Obama and his fellow attorneys demanded that Citibank pay actual damages, compensatory damages, punitive damages, and (of course) attorney's fees.

Citibank denied the charges, arguing that the applicants had been rejected because they had poor credit or were requesting more money than the properties were worth.

As the case dragged on, the parties began to talk about a settlement. Citibank denied any wrongdoing but felt that the expense and length of time necessary to fight the suit would interfere with the conduct of business. By 1997, according to court transcripts, a settlement was near, but "one of the stumbling blocks was the attorneys' fees."[3]

The two sides finally settled in January 1998. (During the 2008 presidential campaign, the media erroneously claimed that Obama and his fellow lawyers had won the case—which is inaccurate.)[4] Citibank agreed to set up credit counseling for applicants who had been denied loans and gave Ms. Buycks-Roberson and two other plaintiffs a total of $60,000. Legal fees totaled $950,000.[5]

Buycks-Roberson was in fact a very small skirmish in an enormous war against American banks and lending institutions, one that had been waged since the 1960s in the name of fighting racism and bigotry. This campaign was based in turn on the charge that banks had used a notorious and manifestly discriminatory practice known as "redlining" to prevent minorities and the poor from getting mortgages and home loans.

Redlining is the practice of writing off entire segments of a city, often black and poor, and refusing to do business there on the grounds that it is simply too risky. Fifty years ago, bank executives would literally draw red lines around sections on a city map and instruct local managers to avoid making loans in those areas. This in turn discouraged real estate agents and construction companies from trying to sell or renovate properties in those neighborhoods. The result was stagnation and deeper descent into poverty, isolating them from any hope of economic recovery.

The federal government, in the wave of 1960s-era civil rights legislation, passed the Fair Housing Act of 1968 in an effort to end the practice. But charges of discrimination persisted, and it has now become an article of faith for American liberals that the mortgage industry is pervasively (if unconsciously) racist.

But was it really true that such practices could be explained by racial prejudice? Or were they in fact rational business practices whose underlying rationale was twisted by skillful demagogues for political purposes?

The question is important, for having ceded this argument to the Left in the 1960s, conservatives have allowed home ownership to be defined as a new civil right, one that must be guaranteed by the federal government, rather than as a privilege to be earned by hard work and wise financial management.

Activists have trotted out reams of statistics to demonstrate what they claim is a clear pattern of discrimination against minorities and the poor. Often they use data which show that when two applicants with the same income apply for the same mortgage, black and Hispanic applicants are much more likely to be rejected. One study by the Federal Reserve Board, for example, found that blacks were twice as likely to be denied mortgages as whites. The consumer advocate Ralph Nader has conducted similar studies and come to the same conclusion. In Texas, an analysis of 400,000 mortgage applicants found that "low income and minority Texans continue to be denied loans at higher rates than the rest of the population."[6] In North Carolina it was found that "middle income black applicants were seven times as likely to have their loans denied as middle income white applicants."[7]

The broad assumption made by the Left and the liberal media is that race—or, more specifically, racism—explains this pattern.

One should not deny that prejudice and discrimination of the kind that flourished under southern segregation may continue in some quarters, even to some extent unconsciously. But one should also beware of attributing all apparent discrimination to irrational racial prejudice. After all, some forms of discrimination are not irrational at all but reflect statistically significant patterns of behavior that strongly correlate with different socioeconomic groups. Such appears to be the case with the redlining studies.

These studies were commissioned in order to provide an objective foundation for allegations of racial discrimination in the banking industry. But there's a problem with the redlining studies: they ignore the important detail of credit histories. None of the studies that have received media attention on this issue dealt with the complexities of loans and why they might have been denied: credit history, outstanding debt, net worth, savings, market value of the home, etc. They simply compare applicants by race and sometimes include their income. But that's not enough to understand why banks make or deny loans to specific applicants.

For banks, after all, it's not just how much money you make that determines whether you get a loan. Studies by the Federal Reserve and the Federal Deposit Insurance Corporation have found that blacks and Latinos tend to have a much higher incidence of troubled credit histories than whites. Indeed, they found that while 49 percent of African Americans have poor credit histories, only 22 percent of whites have the same. Freddie Mac has likewise found that blacks with incomes of $65,000 to $75,000 have on average worse credit histories than whites making under $25,000. This more than anything else explains discrepancies in lending: banks prefer to give loans to people who have a history of paying them back. If you have a poor credit history, you are going to have a hard time getting a mortgage.[8] At least, that is how it used to be in an age of rational banking practices.

But those facts have not gotten in the way of one of the most powerful movements in the history of American politics. Indeed, charges of redlining and racism have proven to be extremely potent weapons. Not only have they allowed liberals and left-wing activists to harass and intimidate political and business leaders, but more subtly and insidiously, they have allowed the Left to argue that unequal results in a wide range of areas—from lending and hiring to college admissions and civil service exams—are manifest proof of "unconscious" or "structural" racism. This intellectual ground, once ceded, could never be regained.

The charge of racism in banking launched a movement in the 1970s that has utterly transformed the American financial system. And this movement, fueled by radical activists, encouraged and supported by liberal politicians in Washington, and funded by Wall Street financiers driven in equal parts by guilt and greed, was the underlying cause of the economic debacle we face today. Since it may take us a generation to dig ourselves out, it may be helpful to understand that it took at least a generation to dig ourselves in.

Seeing banks as inherently racist and discriminatory, liberal activists in and out of government have waged a thirty-year war against our lending institutions. Through their activism, the force of law, and actual or threatened lawsuits, they sought to force banks all over the country to dramatically loosen their mortgage-lending standards.

This movement goes by various names—"fairness in lending," "economic justice," "democratization of capital"—and it has spawned literally hundreds of local housing councils, foundations and alliances. They include—to name a very few—Metropolitan Organizations for People, Latin United Community Center, Massachusetts Affordable Housing Alliance, Fair Lending Coalition, Coalition for Fair Banking, the Maryland Alliance for Responsible Investment, Communities United for Action, and my personal favorite, CA$H PLUS. As we will see, these coalitions are often organized and run by well-known activist groups, civil rights organizations, and labor unions. They represent a grassroots army of the Left.

How successful has the fair housing movement been? By their own account—and they are justifiably proud of their accomplishments—the affordable housing activists and their allies boast of their role in forcing American banks to lend more than *$4 trillion* to people with spotty or questionable credit. Through their actions, literally trillions of dollars have been lent to individuals who would not and should not have been given home mortgages under traditional lending standards. By pushing this agenda, which they pursued with the fervor of a moral crusade, these activist organizations—with crucial cover and support from their liberal allies on Capitol Hill—produced a dramatic reduction of lending standards that ultimately led to an explosion of subprime lending.

As we shall see, charges of racism, bullying threats of obstruction of business, and negative publicity from a compliant press were extremely potent tools that induced banks to play ball, forking over millions of dollars in shaky loans to underqualified minority applicants. The result was a profound distortion of the housing market, artificially driving up prices and pumping enormous amounts of unrecoverable debt into the U.S. financial system.

Yet in the end, when the mortgage bubble burst and housing prices all over the country collapsed, bringing the rest of the economy down with them, these self-same activists outdid themselves in bald-faced hypocrisy by blaming "predatory lenders" for the crisis—as though they themselves had not dragged those very lenders kicking and screaming into the subprime mortgage market in the first place.

Intriguingly, many of the activists who started all this came from Chicago, from the Reverend Jesse Jackson, the founder of Operation PUSH—and a pioneer in shaking down the institutions of corporate America—to an obscure leftist housewife turned activist named Gale Cincotta. As we have seen, Barack Obama played his part as a young lawyer and community activist. The notorious activist umbrella group ACORN played a leading role as well.

Indeed, Chicago was ground zero for the fair housing movement, spreading from there to other reaches of the country. (The very term "redlining" was reputedly coined in Chicago.) And the movement's godfather and patron saint was Saul Alinsky, a legendary rabble-rouser who proclaimed himself a spokesman for the marginalized, poor, and oppressed of America's cities.

Alinsky (1909–1972) made his name in Chicago by organizing the Back of the Yards district (made famous by Upton Sinclair's *The Jungle*) in 1939 and went on to found the Industrial Areas Foundation, a national network of community organizations. His books are considered foundational texts for the activist Left.

He was an odd sort of radical, not what people these days expect a radical to be. There was no long hair, wild clothes, or talk of free love; he didn't tout the virtues of Marx, Mao, or Lenin. He was generally quiet about such matters. Dressed like a fifties-era accountant, with his old-fashioned "square" glasses and short cropped hair, he worked with community groups and churches to advance his agenda.

Alinsky often spoke about "helping the people." But that was not really his goal. He considered such a romantic motivation to be totally naive. Instead he wanted power, pure and simple. A famous story about Alinsky has him asking a group of students why they wanted to be community activists. When they offered platitudes about helping others and making a difference, he quickly stopped them with a raised hand. "You want to organize for *power*!" he screamed. He would proclaim that power is "the very essence of life, the dynamic of life," and should be the ultimate goal.[9]

Alinsky was a radical in the true sense of the word. He despised Lyndon Johnson's Great Society programs, which he considered mere handouts. What Alinsky wanted was not resources but power, which meant controlling resources. In his 1971 book *Rules for Radicals*, he wrote, "*The Prince* was written by Machiavelli for the 'Haves' on how to hold power. *Rules for Radicals* is written for the 'Have-Nots' on how to take it away." He went on to mention the first radical known to man who rebelled against the establishment and did it so effectively that he at least won his own kingdom—Lucifer.

He considered the free-love radicals of the sixties as useless and ineffective; no one would take them seriously. ("They couldn't organize a luncheon.") Moreover, they pointlessly antagonized the middle class. Although Alinsky shared their general contempt for the materialism and selfishness of America's white middle class, he maintained that radicals must work "within the system" and argued that the middle class would have to be brought around slowly to supporting an agenda of radical change. He also grasped the value of a revolutionary crisis:

> Any revolutionary change must be preceded by a passive, affirmative, non-challenging attitude toward change among the mass of our people. They must feel so frustrated, so defeated, so lost, so futureless in the prevailing system that they are willing to let go of the past and change the future. This acceptance is the reformation essential to any revolution.[10]

These words sound hauntingly prescient in light of the current economic meltdown. Obama White House Chief of Staff Rahm Emanuel—Chicago born and bred along with many other members of Obama's inner circle—might have been channeling Alinsky when he said, "Never allow a crisis to go to waste. They are opportunities to do big things." Hillary Clinton—also a native Chicagoan—said the same thing to the European Parliament not long afterward: "Never waste a good crisis." [11]

Alinsky spawned dozens of community organizations in Chicago, and he inspired thousands of activists with his confrontational techniques. Hillary Clinton wrote her undergraduate thesis on Alinsky, and Barack Obama, who was profoundly influenced by Alinsky's disciples, wrote an essay in his honor for a book called *After Alinsky*.[12] The activist organization ACORN—with which Obama has close, long-standing ties—was founded by one of Alinsky's disciples. Another mentored Cesar Chavez of the United Farm Workers of America.

Alinsky understood the world of finance because many of his early backers were financiers. Among the most important was Eugene Meyer, the Wall Street investment banker and chairman of the Federal Reserve from 1930 to 1933. Meyer was also the longtime owner of the *Washington Post*. Other financial backers included Marshall Field III, the department store mogul and an investment banker in New York. Alinsky would brag, "I feel confident that I could persuade a millionaire on a Friday to subsidize a revolution for Saturday out of which he would make a huge profit on Sunday even though he was certain to be executed on Monday."

Alinsky directed much of his fire against banks. He understood not only the power of banks but their potential to act as a fulcrum by which he could advance his ambitious agenda. They also made a convenient enemy. As Alinsky taught, finding an enemy was important. Thus he was quoted in Hillary Clinton's senior thesis: "In order to organize, you must first polarize. People think of controversy as negative; they think consensus is better. But to organize you need a Bull Connor."

When struggling to unionize Chicago meatpackers in the 1930s, Alinsky gave up fighting the companies themselves and went instead to the banks that did business with them. Alinsky threatened them with what he called a "bank-in." He would instruct one thousand workers to come in and open small savings accounts of one dollar each—in effect, shutting down the bank for days. He called it the "savings account tactic." This "could well cause an irrational reaction on the part of the banks which could then be directed against their large customers."

Banks, Alinsky calculated, were the weakest link among his corporate enemies. "If their banks . . . start pressing them . . . they listen and hurt," he once said.[13] Alinsky would use the "bank-in" technique again and again, and his followers would also use it to great effect. Striking steel workers in Pittsburgh employed the tactic in the 1980s, spraying banks with skunk oil and placing dead fish in safe-deposit boxes.

One of those inspired by Alinsky was a Chicago housewife named Gale Cincotta. Small and feisty, she first became active under the tutelage of an Alinsky disciple, Tom Gaudette. Cincotta in turn took up the cause of redlining and racial discrimination and made it her life's work. Embracing Alinsky's confrontational methods, she was once arrested for harassing an administrator at the Department of Housing and Urban Development (HUD) and physically threatened Chicago's tough and imposing Mayor Richard Daley. She once nailed a rat to an alderman's door to protest the city's failure to deal with the rodent problem in her neighborhood (the tactic worked). To raise funds for her activist cadre, she would also shake down and threaten local businesses for donations. (According to her *New York Times* obituary, her favorite "war cry" was "We want it. They've got it. Let's go get it.")[14]

Cincotta, of course, didn't see it that way, declaring that she was simply making use of "hard-sell methods." In fact, she claimed, her tactics were designed to make things easier on businessmen: "We know they are busy and do not have time to donate, so we ask them for money." Local businessmen saw it as something else: a protection racket. Wilbur Gage, the president of Magikist Carpet & Rug Cleaners, told the *Chicago Tribune* that members of Cincotta's organization had picketed his office and warned him to "pay up or get out" because he had refused their request for a $1,500 donation. Gage had given the group $100 before but said he would never do so again.[15]

In the late 1960s, after two people were turned down for loans by a local bank, Cincotta and other members of her community organization discussed the matter with bank officials. When they failed to change their minds, Cincotta and her activists launched a bank-in. The bank quickly caved and negotiated an agreement to offer $4 million worth of loans to the community. Other groups followed suit and organized their own bank-ins in Chicago.[16]

Cincotta lived in Austin, a decaying industrial area of west Chicago. She and her fellow members of the West Side Coalition believed that redlining had kept the neighborhood from being redeveloped, so, with six hundred other activists, she marched on the Chicago City Council. When they were met with shrugs by the hard-boiled council members, they organized a larger protest action involving 1,200 people at the local office of the Department of Housing and Urban Development. In a meeting with federal officials the activists demanded an investigation and got one. The Nixon administration aggressively pushed for seventy indictments against real estate agents, building inspectors, and others. Many of those people were found guilty.

When rumors spread soon afterward that a local bank branch might be closed and reopened in the suburbs, 150 activists held an Alinsky-style bank-in. Some brought their children, who cashed in thousands of pennies. The parents opened savings accounts for one dollar apiece. Others bought money orders for a dollar. They tied up business so badly that the bank president was forced to negotiate. In the end, the bank stayed where it was and the board agreed to invest $3 million in the neighborhood.[17]

These early successes led Cincotta to raise her sights dramatically. This was not just about keeping a local branch open; the goal now was to take decision making away from bankers and financial companies altogether and instead allow local communities to influence how private capital would be invested. Thus, along with a fellow organizer, Shel Trapp, Cincotta founded National People's Action (NPA) and its associated National Training and Information Center (NTIC) in 1972. The activists believed that by "democratizing" capital, they could transform their city and the country. As defined by one advocate, the idea was to "democratize decisions about the distribution of capital by extending at least part of the decision-making franchise to previously 'disenfranchised' people, in particular low-income and minority persons."[18] The effect was to force banks and other mortgage lenders to foot the bill for expanded minority home ownership.

Meanwhile, on Chicago's South Side, a young black activist named Jesse Jackson was coming to the same conclusion. Jackson had already begun to blame redlining as the major cause of black America's economic plight. The affordable housing movement that came out of Chicago ultimately reflected a marriage of Alinsky's confrontational methods and Jackson's patented techniques of racial blackmail.

Born in South Carolina, Jackson played quarterback on the racially integrated University of Illinois football team on scholarship in 1959, only to quit at the end of the season. He would later claim that he had left because of racial discrimination: the white players were not prepared to have a black quarterback. But a later investigation by ESPN revealed one problem with the story: the starting quarterback on the team that year had been African American.[19]

After leaving Illinois he attended North Carolina A&T and joined the civil rights movement in the South. He became a leader with the Southern Christian Leadership Conference and headed up the organization's "Operation Breadbasket" in Chicago. With his gift for public speaking (and rhyming) and flair for drawing attention to his causes, Jackson quickly shook up Chicago with his tactics. After threatening businesses such as the National Tea Company and Del Farm Foods with boycotts, he persuaded company executives to create 370 jobs specifically for blacks and to transfer bank accounts to several black-owned banks. But he feuded with SCLC's leader, Ralph Abernathy, and broke with the group. He then launched the modestly named Operation PUSH—People United to Save Humanity. But there never seemed to be enough money.

In 1971, he wrote to the United Nations asking for donations on the grounds that American blacks were an underdeveloped country. "The United Nations is a great source of support to undeveloped nations and the black population in this country is in effect an underdeveloped nation," he said. The money was intended, of course, for his organization. Alas, U.N. funds were not forthcoming.[20] But you can't fault him for a lack of audacity.

By 1973, Operation PUSH had $6,000 in the bank and debts of $62,000. To fix the problem, Jackson went to friends who ran black-owned businesses and convinced them to make it possible for employees to have donations deducted directly from their paychecks.[21] This was a brilliant innovation, and it opened Jackson's eyes to the power of private businesses to underwrite the costs of social activism. All they needed was the proper incentive. That incentive, he soon discovered, was the subtle (or not so subtle) threat of being publicly charged with racism.

Over the years Jackson developed a finely tuned method for shaking down banks and corporations. First a target company is identified. It needs to be flush with cash and concerned about its image. Jackson makes a public declaration that racial discrimination is occurring in one form or another. Perhaps it is a bank that is failing to lend adequately to the black community. Or perhaps it is a major sporting goods manufacturer that have does not have enough minorities in senior management positions. Whatever its vulnerability, Jackson first calls attention to the problem, alerting the national media, then offers to meet with corporate leaders to help them resolve it. During these meetings he proposes a corrective action as well as a contribution to his organization as a sign of good faith. If they agree—and they usually do if they know what's good for them—a press conference is arranged to announce a resolution of the "crisis." (Note: this does not mean he won't be back for more.) Failure to agree will prompt protests, perhaps civil disobedience, and other techniques designed to embarrass the company.

Like Saul Alinsky, Jackson soon discovered the power of banks. As he would say numerous times over the decades, "Why did Jesse James rob banks? Because that's where the money is." He soon discovered that the redlining argument was a powerful fundraising tool. Indeed, over the course of his very public career as a well-compensated activist, no sector of business would give him more money to leave them alone than the banking industry. Jackson eventually set up a "Wall Street Project," designed ostensibly to secure employment for minorities but more directly intended to pressure banks and investment houses to give him donations. The tactic was brazen and direct, but it yielded rich results. After accusing the New York Stock Exchange of "redlining" minorities, for example, he received a donation of $194,634. Thereafter, his criticisms of the NYSE became more muted.[22] Over the years he received millions in donations from Citigroup, Goldman Sachs, Credit Suisse, First Boston, Morgan Stanley, Paine Webber, and Prudential Securities. He called these donations the "tithe" of corporate America.

In his early Chicago days, Jackson would organize protests and bank-ins in the name of "economic democracy." The Jimmy Carter administration even gave Operation PUSH a $75,000 grant so Jackson could expand his "counseling of inner-city families buying or improving homes in Chicago." Charges of redlining would remain a part of Jackson's verbal repertoire, even when he met foreign leaders. Imagine Japanese Prime Minister Yasuhiro Nakasone's reaction when he agreed to meet with Jackson, only to be lectured about how Japanese corporations were "redlining" blacks in America. Japanese-owned banks were failing to make sufficient loans to the black community, according to Jackson, and Japanese multinationals had failed to hire enough minorities for management positions. Jackson said he would lead a boycott of Japanese products unless a "new relationship" was forthcoming.[23]

Bankers often found themselves in Jackson's sights, and he talked incessantly about redlining. As one columnist for the *ABA Banking Journal* put it, "I often watch 'Both Sides' [Jackson's CNN program] and do not recall ever seeing a segment, regardless of topic and guest list, in which Rev. Jackson did not refer to the problem of widespread bank redlining as one of the main obstacles to progress for minorities and disadvantaged people."[24]

The fair housing campaign took a consequential turn in the mid-1970s, when, after a series of local successes, the activists set their sights on Washington, seeking federal legislation to advance the cause of "democratizing" capitalism. Cincotta, fresh off her string of triumphs in Chicago, took the lead.

In April 1975, Cincotta's group, National People's Action on Housing, sponsored a major conference in Chicago that brought the redlining charge to national attention. She then pushed for—and won—legislation that would require banks to disclose their lending patterns. It seemed like a simple idea: to have lending institutions publicly disclose which neighborhoods they were making loans to and which they were not. Liberals in Congress liked it, and the Home and Mortgage Disclosure Act (HMDA) of 1975 was signed into law by President Gerald Ford. The data that resulted showed pretty much what Cincotta expected: banks tended to lend more in affluent neighborhoods and avoided those that were poor or distressed.

Meanwhile, a series of other studies—conducted mostly by advocacy groups and local newspapers—purported to find racial discrimination rampant in the banking industry. "Black neighborhoods, the studies showed, received far fewer mortgages than mostly white areas did, and black applicants had their loans shot down more often than whites with similar incomes," writes Steven Malanga of the Manhattan Institute. Banks protested that the studies didn't take into account the creditworthiness of these applicants, which was more important than income. "Nevertheless," Malanga observes, "the media worked themselves into a frenzy, pillorying government officials who dared object to the studies' conclusions."[25] As always, the media was only too happy to broadcast sensational charges of racism.

In 1976, Cincotta began pushing for something she called the Community Reinvestment Act (CRA). Again, the idea sounded simple enough: declare that banks have "an affirmative obligation" to lend to people in their own neighborhoods, and make their record of doing so part of the approval process for mergers, acquisitions, or expansion. In short, make it the law that banks needed to lend in areas they had traditionally avoided out of fear that they were poor credit risks.

Cincotta instantly drew support from liberal Democrats in Congress. (What liberal could resist legislation that included the words "community" and "reinvestment"?) The law's biggest champion was Senator William Proxmire, the long-serving Wisconsin Democrat, who immediately put his staff to work with Cincotta's group and other civil rights organizations to draft a bill.[26] Its passage was a major accomplishment and a milestone for the American Left that has not received nearly the attention it deserves.

The law simply said that banks and other lending institutions had an "affirmative obligation" to serve their local communities. There was nothing too disconcerting about that. But what was diabolically brilliant about the legislation were the provisions for how it would be enforced. The law—essentially written by Cincotta and Trapp—left it up to community organizers and activists to monitor local banking activities and empowered them to challenge any significant action that required the approval of federal regulators. If community organizers objected to the proposed bank action, regulators at the Comptroller of the Currency, the Office of Thrift Supervision, the Federal Reserve Board, and the Federal Deposit Insurance Corporation would be tasked with determining whether a given bank was actually serving the needs of local communities, including inner-city areas with high concentrations of the poor and ethnic minorities.

A journalist by trade before entering politics, Senator Proxmire was well known on Capitol Hill for his frugal manner and a certain fanaticism about physical fitness. (He had only a few years earlier published a book titled *You Can Do It: Senator Proxmire's Exercise, Diet, and Relaxation Plan*). He had no background in finance, having spent his whole professional life in newspapers or in government. As a doctrinaire liberal, he completely bought into the redlining narrative because it played to his assumptions about race relations in America. He soon began to complain that banks were engaging in "capital export," whereby banks would actually accept deposits from one area and then make loans to others.

Like other liberals, Proxmire was absolutely convinced that redlining was rampant, and he further assumed that the two practices were closely related, really two sides of one coin: "Banks and savings and loans will take deposits and instead of reinvesting them in that community, they will invest them elsewhere, and they will actually or figuratively draw a red line on a map around areas of their city, sometimes in the older neighborhoods, sometimes ethnic, and sometimes black, but often encompassing a great area of their neighborhood." This would become part of the narrative for every liberal member of the Democratic Party.[27]

One might think that the free flow of capital is a good thing, directing money to the best investment opportunities around the country. But Proxmire was profoundly troubled by it. Speaking on the Senate floor, he complained that banks in Brooklyn invested only 11 percent of their deposits in their own borough. In Washington, D.C., banks made 90 percent of their loans outside the city, and the pattern was repeated in Los Angeles, Chicago, Cleveland, and Saint Louis. The fact that banks were in the business of making profitable loans in order to make money for their owners and shareholders offended Proxmire's moral sensibilities. Banks were making money while "local communities starve," he declared. He was equally blunt about the solution: "Bankers sit right at the heart of our economic system. . . . The record shows we have to do something to nudge them, influence them, persuade them to invest in their community."

But of course, substituting political judgment for economic judgment in banking merely produces another form of discrimination. Lending decisions are no longer made on their economic merits but on the basis of political priorities.[28] It also distorts the mortgage market, with far-reaching effects throughout the economy.

Proxmire was remarkably frank about the political agenda behind the proposed legislation. Even with all of the money being poured into Great Society programs, more than ten years later there was not much progress being made. The federal government, after spending billions of dollars, had little to show for its efforts. And with the economy suffering from inflation, unemployment, and high energy prices, federal funds were not as plentiful as many liberals would have liked. In his remarks to introduce the CRA, Proxmire noted, "Government through tax revenues and public debt cannot and should not provide more than a limited part of the capital required for local housing and economic development needs. . . . The banks and savings and loans have the funds. . . . If we are going to rebuild our cities, it will have to be done with the private institutions."

Putting it that way made it sound like a partnership, but it was more like an offer they couldn't refuse. In effect, Proxmire followed the teaching of Alinsky and Jackson: go to the banks, because that's where the money is.[29]

Proxmire and his allies, who included prominent liberal senators Ted Kennedy, Tip O'Neill, and Frank Church, decided to adopt the path of least resistance and used legislative tactics to avoid congressional debate on the details of the bill. As a result, the CRA, which would have a profound effect on the American economy, was debated in only one chamber (the Senate), and very few legislators were present. Proxmire chose to limit the visibility of the bill and in the Senate Banking, Housing and Urban Affairs Committee attached it to the Housing and Community Development Act, which included the popular Block Grant program, which meant money for senators' home states. The CRA was debated for one day on the Senate floor. In the House there was no debate at all.[30]

Proxmire and his liberal colleagues also made a strategic decision not to frame the bill in the context of race or class—of black and white, rich and poor. Instead, they opted to talk about "places rather than people." The bill was framed as an initiative to promote *community* lending, without actually mentioning which communities they were talking about. As Professor Mary Sidney points out, "Senators avoided discussion of the deservedness of individual beneficiaries that had marked fair housing debates" and instead talked in bland generalities about how the CRA "would contribute to revitalizing declining cities and neighborhoods according to legislators; they did not speak of it as a special program that would channel loans to poor people and minorities."

In this way, they could avoid any discussions about the income or credit histories of those who had been turned down for loans. The senators also avoided talking about the role that community activists would play in enforcing the law. The fact that the likes of Gale Cincotta and Jesse Jackson could testify adversely on the lending habits of local institutions and thereby prevent them from expanding or merging with another bank was not discussed on the Senate floor.[31]

Instead, Proxmire cast the debate in terms of convenience and service. Noting that banks and thrifts were chartered to serve the needs of their communities, Proxmire declared, "convenience and needs does not just mean drive-in teller windows and Christmas Club accounts. It means loans."[32]

There was plenty of opposition to the bill during the short debate in the Senate. Both Republican conservatives and moderate Democrats saw problems with it. Senator Harrison Schmidt, a Republican from New Mexico, worried—quite correctly as it turns out—that the law would "forc[e] financial institutions to make loans of dubious quality." Senator Robert Morgan, a moderate Democrat from North Carolina, called the act "a significant step in the direction of credit allocation by the Congress of the United States." Prophetically enough, he warned, "If bills of this nature are pushed to their ultimate conclusion, then the day will come when a financial institution may be forced to make an unsound loan in a specific location in order to meet its quota of loans in a given locality."[33]

President Jimmy Carter happily signed the law. And it might have been left in obscurity, except for the fact that activist groups now understood that they had a powerful tool to advance their cause.

Nevertheless, the *New York Times* was uncharacteristically skeptical about this bold new piece of social legislation. According to the paper, the Savings Bank Association of New York had concluded that the law "is entirely at variance with the business of banking in a free enterprise system." The association concluded, "Our institutions are not social service organizations." In an editorial the paper remarked that "measures that would weaken standards are dangerous. . . . New York's savings banks already hold large numbers of defaulted mortgages, including many inner city properties . . . we raise a strong word of caution against the expectation that bank credit is a substitute for wages, salaries, and other income that are needed to keep a community alive economically." The editorial concluded, "It is hard to believe that anyone would argue that bad loans are good social investments."[34]

If even the flagship of American liberal journalism was voicing doubts, the potential for mischief must have been glaring indeed. But Cincotta saw nothing but promise in the new law, as did her fellow Chicagoan Jesse Jackson.

So did a young radical named Wade Rathke, another activist who had been inspired by the Alinsky method.

While a student at Williams College, Rathke joined Students for a Democratic Society (SDS) but dropped out of college in 1968 to join the antiwar movement. He then went to work for the National Welfare Rights Organization (NWRO) in Springfield, Massachusetts, organizing the poor and trying to empower welfare recipients. The organizing did not go so well; he could not find enough welfare recipients to forge an effective movement. So he moved to Arkansas and launched the Association of Community Organizations for Reform Now (popularly known as ACORN) along with fellow radicals George Wiley and Gary Delgado of the NWRO.

Rathke and his cohorts openly endorsed the tactics advocated by Richard Cloward and Frances Fox Piven of Columbia University, who argued in *The Nation* magazine that American capitalism could be subverted and ultimately bankrupted by overloading the system with bureaucratic demands. If the state were burdened with welfare payments and other entitlements, Cloward and Piven believed, capitalism would ultimately collapse under their weight. This proposal came to be known in left-wing circles as the Cloward-Piven "crisis strategy."

"We are the majority, forged from all minorities," read ACORN's founding People's Platform. "We are the masses of many, not the forces of few. . . . We will wait no longer for the crumbs at America's door. We will not be meek, but mighty."

ACORN worked to register welfare recipients to vote, fought for a "living wage," and took it on faith that racism and bigotry dominated the financial decisions of banks. The organization benefited early on from federal dollars flowing from the Carter administration. Two ACORN allies, Sam Brown and Marjorie Tabankin, became directors of the VISTA program. (Brown liked to sport a "Robin Hood was right!" button during VISTA board meetings.) And more than $3 million began flowing to ACORN. Over the course of the next thirty years, the organization would grow into an activist behemoth, with at least 350,000 dues-paying member families and more than 800 chapters in more than 100 U.S. cities.

ACORN was profoundly anticapitalist, but it reserved its greatest scorn for banks, which the organization viewed as inherently racist instruments that favored wealthy white patrons over the poor and minority groups. The poor themselves were never at fault if they failed to get loans because of bad credit or lack of income. Instead, ACORN saw credit-scoring methods as inherently racist and "automated underwriting systems," which quantify credit risk, as obviously "discriminatory." ACORN would "blame credit scoring for the limited progress on mortgage lending to minorities," noted the *ABA Banking Journal*. Even math could be discriminatory, in that econometric models used in mortgage underwriting could unduly damage African Americans seeking mortgages.[35]

In the view of the fair housing activists, the fact that urban communities were struggling economically had little to do with crime, poor education, family breakdown, or drugs; instead it was caused by a simple lack of access to money, which was being unreasonably withheld by the banks due to greed, insensitivity, and conscious or unconscious racism. The bankers' argument that they were willing to lend to anyone who could pay back a loan, regardless of color, was met with scorn. These bankers failed to take into account "the unique realities of the poor." For example, banks wanted to give mortgages to individuals who could demonstrate two years of stable employment. For ACORN, this was a ridiculous standard to apply to the poor, who were losing and switching jobs all the time. Banks also wanted credit histories, which demonstrated that when an individual took out a loan he or she had a history of repaying it.

But ACORN and other activist organizations also subscribed to the view that the banks were part of a larger structural effort to keep minorities and the urban poor oppressed. Banks were not simply in the practice of making money; they were political institutions, whether they knew it or not—part of the privileged white power structure. Lending practices therefore were less a function of rational criteria than racial politics. "The banking process is political," as one scholar put it. "Banks decide where credit will flow throughout society and thus what human initiatives will flourish and which will wither. People, ventures, regions, win and lose. This is the stuff of high politics, not calculus." [36]

With the arrival of the CRA, the crusading housing activists discovered that they now had a seat at the table and a vehicle for their redistributionist schemes. By using a cadre of lawyers and an army of activists, ACORN believed it could dramatically alter the terms of the financial free market and force banks to lend to those they would ordinarily not consider qualified. As the sympathetic Professor Heidi Swarts put it, through the CRA, activists could "extract resources" from banks and give them to "poor and working-class people." [37] This is simply a euphemism for extortion.

Another accurate term for the practice would be income redistribution. ACORN could force banks to bend its credit rules and adopt "flexible" interpretations of income and credit history. The CRA would thus allow ACORN and other groups to "redistribute resources from banks to inner-city neighborhoods." [38]

ACORN fought vigorously to force banks to dramatically lower their lending standards, and by the 1990s it was bragging about how it had done away with the old ways of doing business. "At one time a person without formal credit history in the form of credit cards or existing loans was sure to be denied a mortgage," wrote national president Maude Hurd in a 1999 letter to the *Washington Times*. But ACORN had "negotiated groundbreaking Community Reinvestment Act agreements which recognized a history of on-time rent and utility payments as indicators of good credit." [39]

In 1985, ACORN had set up its own housing corporation and forced deals on several large banks that were seeking regulators' approval for mergers. According to ACORN's description of the project in a document titled *To Each Their Home: Success Stories from the ACORN Housing Corporation*, "ACORN Housing and its bank partners developed several innovative strategies. First, they agreed to more flexible underwriting criteria that take into account the realities of lower income communities. For example, income measurements include less traditional income sources such as food stamps, unemployment, part-time jobs, non-court ordered child support and foster care payments. As evidence of creditworthiness, the banks agreed to accept excellent payment histories of rent and utilities in lieu of credit cards, loans or other more traditional forms of credit which many low-income people don't have."

The report also noted that "The banks agreed to lower down payment and closing costs, and to allow family members, churches and ethnic savings clubs to help cover these costs." [40] ACORN Housing even set up its own mortgage brokerage unit with CitiMortgage, Bank of America, First American Title, and Fannie Mae to assist poor and minority families to secure mortgages. In 2004–2005 alone ACORN Housing took in donations close to $400,000 from Citi, $130,000 from the subprime lender Ameriquest Mortgage, $120,000 from Fannie Mae, $1 million from J.P. Morgan, $1.3 million from Bank of America, and

$175,000 from Washington Mutual. It also secured mortgages for undocumented workers and people with undocumented income (meaning income not reported to the IRS), as well as interest-only loans.[41]

To make matters worse, ACORN Housing also receives large grants paid for with taxpayer dollars. It took in $3 million in taxpayer funds in 2004 alone and has received tens of millions of dollars over the decades from the federal government, according to its tax returns.[42]

Not that all of the money ACORN takes in ends up promoting activism. Dale Rathke, a senior official at ACORN and brother of the founder Wade Rathke, embezzled nearly $1 million from the group in 1999 and 2000. When it was discovered, the ACORN board was not notified, nor was law enforcement. Indeed, Rathke remained on the ACORN payroll! The embezzled sum was carried as a "loan" on the books. Rathke and his family repaid some $200,000 of it. He was let go only after the scandal became public. ACORN founder Wade Rathke left soon after.[43]

Those in Congress who might oppose the CRA or attempt to change the law would face the unbridled wrath of the activist cadres. When an aide to Senator Kit Bond of Missouri showed up at a meeting of activists in a church in Saint Louis to discuss his boss's opposition to the law, he was "mauled," in the words of the man who chaired the meeting.[44] In another instance, when Republicans entertained the idea of changing the law in 1991, members of ACORN literally stormed the committee meeting room and brought deliberations to a halt. Several members of ACORN were arrested but were later released when members of Congress appealed to law enforcement to let them go. (According to one report, Congresswoman Maxine Waters refused to leave police headquarters without them.) Such tactics might be good political theater in Chicago, but in the halls of the U.S. Congress they were a shocking breach of decorum that exposed the true face of the activist movement.

ACORN and other activist organizations would also use the law to shake down banks, demanding donations from executives or threatening them with further protests. Lending institutions would be subjected to Alinsky-style bank-ins. But the most effective tool would prove to be litigation. In the case of *Selma S. Buycks-Roberson v. Citibank Federal Savings Bank*, the young Barack Obama was actually serving as an attorney for ACORN. It was one of literally hundreds of suits that the organization would help to file in the name of putting an end to redlining and fighting for social justice.

Some people might dismiss the CRA and the ACORN activists as a distraction. What possible influence could they have on the course of the mighty U.S. economy? Such critics should consider that since 1997, housing activists have forced lending agreements on banks to the tune of $4.2 *trillion* dollars. (To put that into perspective, the entire annual U.S. gross domestic product is just over $14.5 trillion.) The vast majority of these loans were made in the years running up to the financial implosion of 2008. In 2003, for example, $711 billion worth of agreements were signed. In 2004, that number more than doubled to $1.6 trillion. These skyrocketing numbers, staggering to contemplate, come from the National Community Reinvestment Coalition, which bills itself as the "nation's trade association for economic justice" and includes six hundred community organizations that have been active in compelling banks to loan money through the CRA.[45] As we will see, these loans were much more likely to fail than traditional mortgages were, in many instances *three or four times as likely* to lead to foreclosure.

Such mundane considerations, however, are beneath the notice of the radical housing activists. Instead, they are proud of their record of forcing banks to help the poor with nontraditional loans and constantly congratulate themselves as Robin Hoods who take from the rich and give to the poor. In doing so, of course, they are merely acting on their underlying assumption about the sources of wealth. The housing activists, like virtually all leftists, are Marxists who believe that wealth is extracted from the working classes by parasitical institutions such as banks that charge interest for lending money that isn't really theirs. Hence the activists think it right and just to return this expropriated wealth to its rightful owners, and they are prepared to use any means necessary to accomplish their goal. If they break the banks in doing so, that is a small price to pay and may even be considered a positive side effect.

But it is equally important to remember that the funds for these undesirable or shaky loans—better known today as *toxic assets*—do not come from the banks themselves; they come from depositors and investors, people like you and me, while the lost profits that result from loan defaults are absorbed by bank shareholders and by investors in mortgage-backed securities. And as we have learned to our regret in recent years, when banks fail or need to be bailed out, the taxpayers get stuck with the bill.

The housing activists at ACORN and similar organizations saw lending discrimination as a form of class and racial warfare; therefore they saw nothing wrong with using political power to fight what they saw as a systemic form of political injustice. With crucial aid from fellow-traveling liberals such as William Proxmire and Ted Kennedy, they succeeded in pushing through federal legislation that put the power of the state behind their activist agenda. Some years later, of course, the fruits of this socialistic Ponzi scheme would come home to roost, and the culprits would adroitly spin around and blame unregulated capitalism. But from a purely economic point of view, when you use the power of government to force banks to make loans for political reasons, you don't have capitalism. You have a system of organized, state-sponsored extortion.

The resulting devastation of the American banking sector—all in the name of civil rights and social progress—was entirely predictable.

$4 TRILLION
SHAKEDOWN

The Left's Activist Jihad Against American Banks

There's a shrinking pool of funds available, and they're going after you. There's a lot more groups out there than you think, and their knowledge is a lot more sophisticated than you know.

—MICHAEL CRAWFORD, NEW YORK STATE DEPARTMENT OF BANKING

On April 3, 1987, Coleman Young stepped to the podium and surveyed the urban wasteland that was increasingly the city of Detroit. Young, a self-described progressive and onetime member of the Progressive Party, was one of the first African Americans to lead a major U.S. city. But with the auto industry faltering and the city's white middle class fleeing to the suburbs, Detroit was plagued by crime, drugs, and violence. With a clutch of reporters around him, Young lashed out at those he saw as the main culprit in the city's decline: "Our own banks have been notorious for their refusal to invest in the city, to extend loans to people who would invest in the city, or even to give mortgages to people who want to buy homes in the city," he declaimed in an angry voice. "That's no secret. . . . These banks got fat and got rich off the City of Detroit, but you know, this is not a sentimental game. Ain't no such thing as gratitude in finance."[1]

The 1980s might have been the Ronald Reagan years in Washington, a time of vigorous capitalism and commitment to free-market verities, but around the country hundreds of activist groups and urban leaders, empowered by the Community Reinvestment Act and the lure of racial and class politics, were bent on politicizing the banking system. The solution to their problems, they believed, lay in forcing lending institutions to make risky loans in urban areas and set aside funds for selected socioeconomic or racial groups. Egged on by a media with an appetite for stories about racism, class warfare, and rising income disparities, the activists would increasingly demand a say in how mortgage loans were made. Using fear and intimidation and the megaphone of a sympathetic press, they would begin to chip away at lending standards, weaken underwriting rules, and push banks away from their traditionally conservative practices.

Some of the organizations fighting this battle tried to use reason to make their case. "Confrontation is not our style," said Larry Swift of the Chicago-based Woodstock Institute. "It's better to keep the club under the table."[2] But most preferred confrontation because it seemed to get better results. Alinsky, it seems, was correct: banks would usually back down in the face of controversy and attack.

New groups got involved in the campaign. The Center for Community Change became enmeshed in the shakedown agenda and was helpful in pushing the cause. Founded in 1968, the CCC was committed to the Alinskyite vision and boasted a board of directors that over the years included Peter Edelman, the husband of the civil rights icon Marian Wright Edelman; Congressman Ron Dellums; and Paul Booth, a founder of Students for a Democratic Society. As it fought to weaken banking standards, the organization was heavily funded by the Fannie Mae Foundation, the Rockefeller Foundation, the Carnegie Corporation, and the Woods Foundation of Chicago.

Other national organizations saw the power that the CRA would bring them and put their considerable resources behind these efforts. Labor unions such as the United Mine Workers, United Autoworkers, Service Employees International, and Amalgamated Clothing and Textile Workers Union became involved in a national campaign to shake down banks. And civil rights organizations such as the NAACP and the Southern Christian Leadership Conference joined Jesse Jackson in the cause. National activists such as Ralph Nader also mobilized their supporters.

This national movement would have a dramatic—and ultimately traumatic—effect on the U.S. economy. Yet the story of what these activists did and what they accomplished has never before been told.

In little more than half a dozen years, the campaign would be able to claim some notable successes. By 1987, newspapers were filled with headlines about how activist groups could "instill fear in banking circles" and were forcing "more dollars . . . into troubled neighborhoods." The *Washington Post* columnist Neal Peirce would report that there were bank challenges (aka shakedowns) taking place in nearly 120 cities in 39 states, with billions of dollars at stake. *USA Today* described Gale Cincotta "as a woman who makes bankers quake in their pinstripes" and faithfully reported her antics. When she invaded the Washington apartment building where Housing and Urban Development Secretary Samuel Pierce lived and stuffed fliers with his picture under every door that read "Housing Enemy No. 1," she was portrayed as a folk hero. "This is our style," the paper quoted her as saying. "If we're uncomfortable, we'll make you uncomfortable until you do something."[3]

Jesse Jackson raised tens of millions of dollars in the name of taking it to the banks and ran for president twice on a platform that demanded banks invest more money in poor neighborhoods. He made access to loans and mortgages a civil rights issue and condemned those who defended traditional lending practices as "racist." The Reagan decade also brought about the rise to national prominence of ACORN, with its radical tactics and its willingness to physically intimidate bankers, politicians, and anyone else who stood in its way. Accused of "Trotskyite" tactics, ACORN stoked great fear in bank boardrooms. "ACORN has passed Gale Cincotta in instilling fear in banking circles," explained Frederick Manning of the Philadelphia Fed in 1987.[4]

Just as its framers had envisioned, the Community Reinvestment Act allowed activists to challenge bank mergers, obstruct the opening and closing of branches, and even block the introduction of ATMs on the grounds that they were failing to meet the needs of the local community. Often, these activists didn't even live in the communities they were supposedly defending.

In 1986, for instance, Fidelity Bank of Philadelphia was planning to buy out another bank but soon found ACORN standing in the way. In order to clear the radical group's objections, Fidelity agreed to ACORN's demand that it count food stamps as income for lending purposes. This meant that suddenly people who previously hadn't qualified could get a house and a mortgage. Typical was the case of Joan Arrington, who had applied for a loan with Mellon Bank a year earlier. Unemployed and on a partial medical disability, she had been turned down by Mellon because she lacked clear title to the property. "When you're poor, the establishment puts a foot across the back of your neck and they refuse to take it off," she complained to a local paper. But under the agreement with ACORN, Fidelity now gave her the mortgage.[5]

In 1989, Chase Manhattan was forced to commit $200 million to New York City's poor neighborhoods after it was charged with "racially discriminatory practices" by ACORN and Operation PUSH.[6] ACORN and the United Mine Workers of America filed a thirty-one-page complaint with the Federal Reserve, claiming that the bank had "discriminated against minority and low-income neighborhoods in Brooklyn."[7] How on earth did the United Mine Workers get involved in a case involving lending in New York City? Simple. It was trying to pressure Chase Manhattan, which was the banker for mine companies it was negotiating with. Pure Alinsky.

Two years earlier, community activists had challenged the efforts of Chemical Bank to open a new branch in their Bronx neighborhood on the ground that the bank needed to make more loans to the poor and minorities. Chemical eventually succumbed to the pressure and offered to provide $25 million in loans at below-market rates.[8]

What it couldn't get, ACORN simply took. In 1985, the organization illegally seized twenty-five abandoned buildings in New York City and moved in squatters. Eleven people were arrested. Nevertheless, City Hall eventually caved in and gave the group title to all the buildings. The group then tried to force lenders to refurbish the property they had seized.[9]

ACORN was anything but shy about its tactics. While shaking down the Bank of New York in 1988, the group accepted a $250,000 interest-free loan and a deposit of $100,000 in an ACORN account just to get it to go away.[10]

New Jersey was the site of the shakedown of Carteret Bancorp, which was set to merge with a New York lender. A coalition of groups, including Ralph Nader's Citizen Action, the NAACP, and other housing groups, fought the application. In the end, the company had to cough up $52.5 million in mortgages for lower-income families, which were offered at below-market rates. It also agreed to fund construction loans to nonprofit groups that wanted to build affordable housing. The loans were made at prime rate with no fees.[11]

In Albany, New York the state banking department reported being "deluged" with requests for information about how the act could be used by "a spectrum of special interest groups." According to Michael

Crawford, the director of the department's CRA Monitoring Unit, the groups wanted to know how it could "leverage" more money out of the banking industry. He told a group of bank executives in Syracuse, "There's a shrinking pool of funds available, and they're going after you. There's a lot more groups out there than you think, and their knowledge is a lot more sophisticated than you know." [12]

In Massachusetts, a group with the Orwellian-sounding name Massachusetts Fair Share joined with other groups to go after Capitol Bank and Trust of Boston on the grounds that it was not lending enough to the poor black neighborhood of Roxbury. The bank wasn't making loans there, "but it's not because there is a shortage of people in Roxbury seeking mortgages," as one activist put it. Whether there was a shortage of people with good credit histories in Roxbury he didn't say. [13]

But the group had reason to anticipate success, since a few years earlier, the New England Merchants National Bank, seeking a merger with Massachusetts Bay Bancorp, had agreed to set aside 25 percent of its mortgage funds for low-income and minority applicants. The bank was being rolled by the Massachusetts Urban Reinvestment Advisory Group (MURAG), which had the active support of liberal Massachusetts regulators. [14] Other banks, such as the Provident, had their applications to open new branches denied because they would not cave in and sign such agreements. [15] The warning did not go unheeded.

Another coalition of Massachusetts liberals went after the Somerset Savings Bank, which wanted to create a holding company. State regulators needed to approve the move, and they forced the small bank to cough up $25 million in loans. "We didn't want to fight it and drag out the application process," said bank president Thomas Kelly, sounding as if he were negotiating the freeing of hostages. "It was a no-win situation." [16]

There were plenty of other banks that took Kelly's view and

avoided a fight altogether, quietly cutting deals that are not even publicly known. As one supporter of these actions notes, "When examining challenge activity, it is important to keep in mind that, for every formal CRA challenge, there were many CRA negotiations that resulted in CRA agreements or activity before any formal challenge was filed." Merely the threat of such action "prompted banks to proactively seek relationships." Relationships? Perhaps the kind that exists when a robber puts a gun to your head.[17]

In Chicago, Gale Cincotta forged an unprecedented coalition of sixty community groups and put three large banks in her sights: First National Bank of Chicago, Northern Trust, and Harris Trust. She squeezed out a commitment for $153 million in loans but warned that it was "nowhere near" what was needed. She would be back for more.[18]

When Continental Bank of Chicago sought to buy a small bank in Arizona, the Amalgamated Clothing and Textile Workers Union and ACORN fought it on the grounds that Continental was not complying with the CRA. The Federal Reserve actually rejected the merger on the ground that the bank had failed to meet the credit needs of Chicago neighborhoods.

Cincotta's organizations also trained up a new generation of activists. In the mid-1980s a young college graduate named Barack Obama went to the NTIC to learn the essentials of community organizing. The sessions were held on the South Side of Chicago and run by Cincotta's partner, Shel Trapp. Obama would spend four years heading up the Developing Communities Project, which received assistance from Cincotta's organizations and employed Alinsky-style tactics to mobilize the black community.

Obama embraced the activist agenda of shaking down banks and forged a close relationship with activists at ACORN. One of his allies was Madeline Talbott, an organizer who was leading the ACORN campaign against Chicago-area banks in favor of the CRA. After he returned from Harvard Law School in the 1990s, Obama would conduct classes for future leaders identified by ACORN. When he ran for the Illinois State Senate in 1995, he explained that his goal was to "stand politics on its head" by empowering citizens and forging a union among "banks, scornful grandmothers and angry young." Talbott called him "a kindred spirit." [19]

Nationally, the campaign gained intensity. In June 1989, protestors starting singing and waving placards in front of the First Union Corp. of North Carolina, demanding that the company contribute to a fund for low-income neighborhoods. [20] Edward Crutchfield, the chairman and CEO of First Union, found the activists to be "a pain in the neck as far as I'm concerned." They practiced "forms of pure blackmail," he said. Vice chairman Tom Rideout told the American Bankers Association that the campaign was nothing short of an "effort to turn banks into public utilities." [21] But First Union would eventually relent, despite the protestations of its senior executives, cutting deals in both North Carolina and Florida so that proposed expansions could take place. The outcome seemed inevitable, however, given that three years earlier Legal Services of North Carolina had successfully pressured First Union to begin conducting "innovative home mortgage lending" to poor and minority applicants in order to complete a merger with a small Georgia-based bank. [22]

Plenty of voices raised concerns about these practices. As early as 1981, Richard T. Pratt, the chairman of the Federal Home Loan Bank Board, said the policy was detrimental to the "safety and soundness" of America's lending institutions. He worried that government regulators were in effect trying to allocate credit. In a free market, credit is allocated by the market itself, not the government. [23]

In 1982, the Federal Reserve Board confirmed that the CRA was creating a situation in which "credit allocation" was taking place. Banks were not making lending decisions based on sound business principles; rather, they were increasingly making decisions based on social and political pressure. The Fed report noted that at that point the commitments were not so great: "To date this credit allocation has not been of such magnitude nor has it imposed such severe constraints that it has threatened an institution's basic safety or soundness."[24] Then again, the activists were just getting started.

Down in Louisiana, where ACORN had its national headquarters, the organization went after Hibernia, which was attempting to merge with a small bank in Lafayette, Louisiana. ACORN, joined by the powerful Service Employees International Union (SEIU), filed a protest with federal regulators and began picketing outside Hibernia branches. The activists demanded that the bank develop a quota system for offering mortgages to poor and minority applicants. They wanted Hibernia to "consider all wages, including public assistance and food stamps, when determining credit-worthiness." According to the bank's general counsel, ACORN even demanded that the bank give it a $50,000-a-year grant. The bank refused and fought back, much to the consternation of ACORN. "Banking is a business," said bank counsel Bob Coury. Even so, Hibernia was forced by federal regulators to develop a comprehensive plan to offer more mortgage products to these groups in order to get the merger approved.[25]

When Continental Bank of Philadelphia sought to merge with Midlantic Bank of New Jersey, ACORN threatened to protest unless Continental gave it what it wanted. Activists treated bankers to a "power lunch" at a Camden, New Jersey, soup kitchen and threw up picket lines at several Midlantic locations. Continental was quick to capitulate. The bank agreed to "broaden its lending policies" and to "loosen its criteria for mortgage loans" and offer "lower interest rates for and require less equity for loans." It would provide "below-market rates on home mortgages" and accept "sweat equity in lieu of cash down payments."

In keeping with ACORN's agenda, Continental agreed, as so many others were now doing, to include "income from sources not traditionally considered, such as food stamps, part-time jobs and seasonal employment" in judging creditworthiness. In changing a policy that had been in place from the beginning, Continental also stipulated that "lack of a credit history will not be sufficient to stop a mortgage loan." ACORN leader LaVerne Butts called the agreement "a major step toward getting money into the lands of low- and moderate-income people."[26]

By the mid-1980s, activist organizations around the country had learned that money could be squeezed from lenders anxious to avoid bad publicity. "Every year we probably see about twice as many protests to bank mergers or charters than the previous year," said Hugh Loftus, a bank vice president in Seattle. Accusations of discrimination "are a fact of life, and banks are adopting procedures to respond to it," he told colleagues at a seminar. Loftus's Rainier Bank quickly found itself the target of protestors, who were demanding that minorities in depressed neighborhoods be given "special, lower standards on bank loans." The campaign was organized by Keith Dublanica, a bus driver for Seattle Metro who said he was incensed by the discrimination he saw around him. He organized the South End Seattle Community Organization (SESCO) and filed a protest with the Federal Reserve Bank against the sale of Rainier to California-based Security Pacific. Dublanica argued that Rainier was discriminating against minorities by not offering loans with flexible lending criteria. What he wanted, explained the *Seattle Times*, was a "kind of affirmative-action proposal in the making of loans." Poor minority neighborhoods were not receiving enough in loans, so Dublanica had a simple solution: offer a 2 percent discount on rates and waive fees for certain (minority) neighborhoods. As he put it, "If the racial mix is different from the rest of the city, perhaps the criteria should be somewhat relaxed." Though Security Pacific resisted the blatant quota system Dublanica wanted, it did end up putting aside $2.4 billion for special loans.[27]

"Big banks all over Florida are running a little scared," warned the *Miami Herald*. "It's not bigger banks they fear. It's a ragtag group of young, public-minded lawyers who have turned a little-used law into a powerful weapon." The article explained how "shaggy haired lawyers," many of them working for Legal Services or other nonprofit activist groups, were targeting banks through the CRA. The *Herald* noted that many banks "wonder how they can lend money to people who can't pay it back and why profit-seeking institutions should be responsible for the economic development of poor communities." But the activists themselves didn't seem interested in those questions. "We're not saying they should open the vaults to everybody in Liberty City [a poor black neighborhood] and let them in," said one activist lawyer. "But if the CRA is the only way to get banks working with low-income people, then I say so be it." [28]

The lawyer failed to note that the CRA had been drafted by the activists themselves and was intended to be used in precisely this way.

Legal Services lawyers forced banks in a series of agreements to accept "flexible credit requirements" in making loans. Much of the legal heavy lifting was done by Jay Rose, a conga drummer in a reggae band and legal aid lawyer in Orlando. Rose, a youthful radical who in high school had mockingly changed the last phrase of the Pledge of Allegiance to "With liberty and justice for some," was branded a "white social dilettante" by Orlando Mayor Bill Frederick. "I believe human rights are more important than property rights," he told one reporter.[29] But Rose got results and announced that the successful campaign "will set the tone for a whole new philosophy of banking in the state." [30] In Dade County, a grand jury in a lending case actually recommended that Metro-Dade and Miami consider depositing half of their combined $35 million in federal revenue into a black-owned bank.[31]

Barnett Banks of Florida, then the state's largest lender, was challenged by the taxpayer-funded Florida Legal Services for its failure to lend enough to poor and minorities. Legal Services said the "banks could do more in low-income areas," and Barnett, not wanting a fight, agreed to set up a "special loan program" for borrowers, including flexible underwriting criteria.[32]

In Atlanta, Reverend Joseph E. Lowery of the Southern Christian Leadership Conference tried to block Trust Company Bank from establishing a new branch in Cherokee County, complaining that under the CRA the bank didn't make enough loans to minorities. He also criticized the company for not having any black members on its board of directors.[33]

Shortly afterward, Cleveland's Society Bank started offering 5 percent down payment home mortgages with $1,000 grants to poor borrowers.[34] In San Antonio, ACORN and Gale Cincotta's National Training and Information Center went after six banks that were seeking to merge or expand their operations. The activists succeeded in getting the banks to lower their lending standards and obtained an agreement whereby their groups would actually screen the applicants. "The quid pro quo is that these groups will act as our marketing arm," explained an official with First Republic.[35] A novel approach to "marketing," to be sure.

In Dallas, ACORN went after Allied Bancshares, which was due to be acquired by First Interstate Bancorp of California. ACORN demanded that the bank make more loans to poor and minority applicants. "It's our money and we want it back," declared Arquilla Smith, the chairperson of ACORN in Dallas.

Smith's pronouncement made explicit the unstated principle at the core of the activist campaign: the money in question did not belong to the banks themselves or to their depositors or shareholders but to society at large. Call it the Robin Hood principle, founded on the Marxist premise that all accumulated wealth is ipso facto an unjust expropriation of collective resources.

In Saint Louis, ACORN went after Centerre Bancorporation, accusing it of redlining parts of the city. The bank relented and agreed to make more loans available to the poor. It agreed to loosen lending standards and "to lend up to 97% of a house's appraised value," noted the *St. Louis Post-Dispatch*, adding, "Usually, banks require down payments of 10 percent or more on mortgages." Boatmen's Bancshares in Saint Louis struck a similar agreement with ACORN, totaling $70 million.[36]

California First Bank, also on the hit list, agreed to $84 million in special loans.[37]

Just because a bank happened to be black-owned bank didn't mean that it was immune to such tactics. If it maintained conventional lending standards and failed to make the specified loans requested by activists, even a black-owned bank could face charges of racism. When Independence Bank of Chicago wanted to acquire another bank in the city, hoping to become the largest black-owned financial institution in the country, the South Side Community Development Coalition declared that the bank had failed to offer enough mortgages to minorities. (Ironically, Independence Bank of Chicago was run by allies of the Reverend Jesse Jackson. Its vice president, Cirilo McSween, was Jackson's financial adviser and a founding board member of Operation PUSH. For some reason, Jackson didn't join this protest.) Activists went to federal regulators in an attempt to block the move, and Independence eventually agreed with activists to make $7 million in loans to applicants with poor credit. The same thing happened to Chicago's black-owned Beverly Bank, which signed a $20 million lending pact with the activists.

Chicago's Highland Community Bank, another black-owned bank, had made plenty of loans on Chicago's South Side, but activists demanded more. Highland's president groused that the Community Reinvestment Act "is a tremendous amount of power to put in the hands of people who are not sophisticated about banking." He refused to even meet with the activists.[38]

In Michigan, the black-owned Omnibank was the target of a

shakedown by activists and accused of racism, but bank chairman William Johnson fought back. He declared that the CRA was "a flawed social policy which has proven to be economically counter-productive." Noting that the law tended to favor black borrowers over whites, he charged that the CRA was "race-based legislation" and expressed the hope that it would be overturned by the U.S. Supreme Court. But there was no such luck for Mr. Johnson. The U.S. Supreme Court has never ruled on the constitutionality of the CRA.[39]

The activists gained even more momentum with the help of friendly journalists. Newspapers simply could not resist a sensational story about racism, greed, and heartless bankers. A predictable pattern emerged: banks that caved in to the activists would be granted wonderful media coverage, lauded as progressive and compassionate. Those that resisted would be labeled heartless obstructionists. In New England, for instance, when twenty-five regional banks signed lending agreements, the *Boston Globe* called them "banks with 'white hats.'"[40]

The *Atlanta Journal-Constitution* won a Pulitzer Prize for demonstrating, based on federal data, that between 1983 and 1988 the rate of rejection on mortgage applications in the Atlanta area had been 11.1 percent for whites while the rate for blacks was 23.7 percent and that for Hispanics 18.2 percent. The newspaper series created a firestorm in the city, leading ministers and community leaders to stage protests and call for the mass withdrawal of deposits from banks named in the article. The series garnered national attention when Congressman Joseph Kennedy railed that antidiscrimination laws were being "pervasively disregarded" by those banks and (by implication) others across the country. In a grandstanding speech on the House floor, Kennedy declared that "every American citizen" should have "equal opportunity to secure credit in order to own a home."

Ignored by everyone, of course, was the fact that the data did not take into account such things as income, debt burdens, credit histories, or net assets of those whose applications had been rejected. The attention-getting charge of racism was simply too appealing to the media, the activists, and liberal politicians like Kennedy. When pressed by critics, Kennedy dismissed these objections: "I don't believe anybody with reasonable judgment would come to that conclusion."[41]

The Atlanta banks, bullied into compliance with the activist agenda, immediately offered to alter their lending standards in order to eliminate their allegedly racist practices.

Not to be outdone, the *Detroit Free Press* ran a four-part series imaginatively called "The Race for Money." Using the same national data source as the *Atlanta Journal-Constitution* but applying it to Detroit-area banks and applicants, the series charged that banks avoided lending to blacks and favored whites when it came to home mortgages. Redlining in its traditional form was no longer being practiced, the paper concluded; instead, "a more subtle process of discrimination" was taking place. The series explained that loans were given three times more often in white census tracts than in comparable black neighborhoods. The newspaper also alleged that Detroit lenders were reluctant to participate in government programs designed to promote loans to those with troubled credit histories.[42]

As in Atlanta, the series sparked outrage as activists turned out in full force, and the local government was abuzz with threats of legal action against the problem banks. Among activist groups it was a call to outright war. A coalition was quickly put together, including the United Auto Workers and the NAACP, which voted to "negotiate" for more loans "or force it." UAW vice president Marc Stepp outlined the group's strategy: "Pick one, maybe the worst one or two banks, and level correction action demands. And if that doesn't work, level punitive demands. That will work. Take the money out. Shut them down. That will work." Alinsky himself could not have said it better.[43]

Once again, elected officials in Washington couldn't help but draw the wrong conclusions. Senator William Proxmire, the father of the CRA, demanded that federal regulators examine the newspapers' findings. "These," he said, "are alarming allegations—particularly the charge that racial discrimination in home lending persists." [44]

If in Atlanta the banks capitulated, those in Detroit would not go down without a fight. Activists quickly settled on attacking Comerica, which had been mentioned in the *Detroit Free Press* story. The Michigan-based bank was interested in acquiring a bank in Texas, which would require the approval of regulators. Comerica hired David Goldberg, a professor of sociology at the University of Michigan, to analyze the newspapers' research. While noting that the data did not include critical information like the credit histories of those who had been denied mortgages, Goldberg went even further in concluding that the research was flawed and highly deceptive. Using regression analysis, he found that the charge of racism was baseless.

"Comerica does not practice discrimination in mortgage lending," he concluded. The differences in lending rates between blacks and whites were caused by legitimate financial factors. "No honest person could conclude that there was discrimination in mortgage lending" based on the numbers, said Goldberg. The only reason one would make such a charge in this case is because of "mindless, knee-jerk reaction to endless media bombardment," motivated by the desire of journalists to win awards and embrace a "Big Lie." [45]

But facts are easy to ignore when there are large sums of money at stake. Under pressure from activists, Michigan regulators determined that Comerica could not complete its acquisition until it submitted a three-year plan to increase lending in inner-city Detroit. Comerica, desperately wanting to expand its operations during a time of merger mania in American banking, knuckled under and cut a deal with the activists. The bank agreed to cough up $280 million for minority loans in Detroit and also agreed to "broader underwriting policies" in order to make those loans easier to get.

Comerica got off easy. The National Bank of Detroit, also targeted by activists, pledged more than $1.9 billion to the cause. In all, the *Detroit Free Press* newspaper series and the outrage that followed led to more than $2.9 billion in loans to unqualified borrowers. The amounts just kept going up and up.[46]

The aggressive push for CRA loans naturally created tension within banks. The simple fact is that getting a good CRA rating increases problems with safety and soundness. Jeffery Gunther of the Federal Reserve Bank of Dallas statistically analyzed 16,212 sets of ratings for the CRA and the so-called CAMELS rating used by the federal government to determine the soundness of banks. What Gunther found is that the better a bank's score on the CRA, the lower its CAMELS rating and vice versa.[47]

In the 1980s, activist groups began forcing banks to erode their lending standards in the name of social justice. Many of their demands had embedded in them the very lending practices that would lead to the explosion of subprime lending and the consequent financial meltdown. Banks could get CRA credit for issuing so-called subprime loans.[48] And the agreements negotiated with the activists included, for the first time in banking, no-money-down loans, loans to people without jobs, and loans larger than had previously been allowed. Banks traditionally did not want housing to account for more than 28 percent of a borrower's monthly income. Activists argued that "these standards in many instances are too low for low-income and minority individuals." They succeeded in forcing many banks to raise that number to 33 percent and in some cases 40 percent.[49]

The CRA applied only to banks; it did not apply to mortgage lenders such as Countrywide or Ameriquest. But the CRA laid the foundation for troubles to come. As Peter Wallison, the Arthur F. Burns Fellow at the American Enterprise Institute, put it, "The most important fact about the CRA is the associated effort to reduce underwriting standards so that more low-income people could purchase homes. Once these standards were relaxed . . . they spread rapidly to the prime market and to subprime markets, where loans were made by lenders other than banks."[50]

There is a direct line between the CRA and the explosion of so-called subprime loans a decade later. As Alan Greenspan explained in testimony before the House Committee on Oversight and Government Reform in October 2008, "It's instructive to go back to the early stages of the subprime market, which has essentially emerged out of the CRA."[51]

Those sentiments were echoed by Federal Reserve Board Governor Edward Gramlich, who in a speech discussing the origins of subprime lending argued that the "evolutionary process" of such loans had been helped substantially by the CRA, which "gave banking institutions a strong incentive to make loans to low- and moderate-borrowers or areas."[52]

Those efforts opened the floodgates to loose lending standards and creative financing such as no-money-down loans. Evidence was just starting to accumulate that these loans had failure rates as much as *seven times* as high as ordinary mortgages. But of course, for the activists and their allies in Washington, that was hardly the point. They didn't care if the loans were repaid, only that wealth was redistributed from the "predatory" banks of their inflamed imagination to the deserving poor, to whom the money "really" belonged.

Over the course of little more than a decade, the activists had accomplished much. By waging a jihad against traditional lending standards, they had helped to spark the growth of subprime lending. But their efforts were about to be turbocharged as a generation of guilt-ridden baby boomers assumed power in Washington, D.C., and put the full weight of the federal government behind their campaign.

THE CLINTON CRUSADE

How Democrats Made Credit a Civil Right

How in hell did we qualify?

—VICTOR RAMIREZ, A STUDENT WITH A SALARY
OF $17,000 AND MORTGAGE RECIPIENT

For more than fifteen years, fair housing activists had been using the Community Reinvestment Act to compel banks to make increasingly risky loans. Using tactics of intimidation, delay, and public embarrassment, they had achieved stunning results. By 1990, some $5 billion had been shaken from banks through these tactics. But with the election of Bill Clinton in 1992, what had been a nuisance and a low-level operation against banks would become a full-fledged effort to use the power of the state to shape the lending policies of banks, bringing them into line with the activist housing agenda.

From the beginning of his candidacy, Clinton had made home ownership for the poor and minorities a centerpiece of his urban policy. He mentioned the Community Reinvestment Act more than any other presidential candidate before or since.[1] And with his election in 1992, activists discovered that they were no longer simply nipping at the heels of the financial establishment; they now had a friend in the White House who could dictate to that establishment. Clinton believed that millions of Americans were being held back by banks with too-stringent lending practices. This was manifestly (to him) not being done simply in the name of business soundness but was prima facie evidence of bigotry.

First Lady and political partner Hillary Clinton was also a strong ally. Not only did she embrace the agenda of spreading minority home ownership, she supported the Alinsky method of achieving it. During her time in the White House, she would raise money, attend events organized by an Alinsky affiliate, and lend her name to projects endorsed by Alinsky's Industrial Areas Foundation. The first lady also met with Alinsky organizers in the White House several times.[2] This is hardly surprising, given her long-standing interest in community activism.

The Clintons would not simply pay lip service to these efforts. With the trademark hubris and impatience of his generation of liberal baby boomers, President Clinton would embark on a massive social engineering program that would, in the hallowed name of civil rights, dramatically undermine the lending standards of banks all over the country. He thereby set into motion a series of events that would shake the financial foundations of the country—and the world—sixteen years later.

Clinton and his team were part of a new wave of liberal activists turned political careerists who would rise to power with him. Highly educated technocrats—products of the "moderate" wing of the 1960s revolution who chose to work to change the system from within rather than seeking to destroy it—they also displayed a unique ability to square their raging personal ambition with high-minded social ideals, all the while excusing or overlooking their own hypocrisies and occasional venal sins.

These were not at all the same type of people as the community housing activists. Gale Cincotta, the tough-as-nails working-class Chicago housewife, would have had little in common with the Ivy League–educated members of Clinton's inner circle. While Cincotta lived in the midst of the struggles of Chicago, members of the Clinton inner circle revolved into and out of lofty positions in government, business, and academia. With their elegant town houses and chauffeur-driven lifestyles, they wanted for very little.

What they did desire was the moral clarity and urgency of the civil rights movement, which had been the defining issue of their generation. Liberal boomers such as the Clintons and their friends had an abiding nostalgia for the drama and passion of the civil rights movement, and many were racked with guilt because they had either missed out on it or failed to participate. But they were also unwilling to give up their comforts or abandon the path of financial success. So they created a hybrid form of activism that would allow them to pursue their own goals while claiming the civil rights movement as their own by embracing the fair housing agenda. To this end, Clinton officials at HUD and DOJ teamed up with local activists to put the squeeze on U.S. banks.

Typical of the new breed of boomer liberal was Robert Rubin, who would serve as Clinton's Treasury secretary. Not only would he be a major architect of Clinton's economic policies, he would serve as a mentor to many of those who are now key members of President Barack Obama's economic team.

Rubin grew up in Miami as the son of a well-to-do attorney. He attended a Jewish temple, but it was largely a formality (the rabbi once told Rubin that he didn't believe in God in a conventional sense). In a memoir, Rubin recalled that as a youth he was aware of racial discrimination in Miami and of the "immense injustice being done to so many."

Rubin went on to study economics at Harvard, but the biggest influence on his life was a philosophy professor named Raphael Demos, from whom he learned that we live "in a world without absolutes or provable certainties." From Harvard, he went to the London School of Economics, then attended Yale Law School. After a brief stint at the Manhattan law firm Cleary Gottlieb, he joined the arbitrage department at Goldman Sachs, where he spent the next twenty-six years. (Among his colleagues was future New Jersey Senator and Governor Jon Corzine.)[3] In 1993, he joined the Clinton White House as the director of the newly created National Economic Council. Two years later, after Clinton's reelection, he was sworn in as Treasury secretary. Rubin will play a major role in our story, largely behind the scenes.

Clinton himself came by his interest in civil rights honestly enough. He had grown up in the South and knew firsthand the face of black rural poverty. He sincerely believed that increasing home ownership among the poor, blacks, and Hispanics would solve a host of social problems and be good for the country as a whole. There would be less violent crime, higher test scores, stronger communities, even less graffiti in poor neighborhoods. These were laudable goals, of course, and his views were shared by a great many people in Washington, both Democrats and Republicans. But his understanding of the causes of the problem—and therefore the solution—were fundamentally flawed.

Like other liberals of his generation, Clinton took it as an article of faith that the biggest impediments to minority home ownership were the conservative lending practices of banks. These invariably involved an examination of credit history, proof of stable income, evaluation of net worth, the size of the down payment, and the loan-to-value ratio. Given those stringent standards, many who deserved to own homes (in Clinton's view) could never get them. As governor of Arkansas, he had tried to address the problem by establishing the so-called Good Faith Fund. This was a lending institution that would offer loans to the poor to help them start businesses and improve their homes. Unfortunately, when repayment rates fell to only 48 percent after just two years, the fund went out of business.[4]

Undeterred by that experience, Clinton made fair housing a major commitment of his new administration. He set the tone early, mentioning the CRA during his first State of the Union Address. "America will never be complete in its renewal until everyone shares in its bounty," he declared to thunderous applause from Democrats.

But as the economist John H. Makin pointed out in an illuminating article, Clinton took Democratic housing policy in a radically new direction. Previous Democratic administrations had sought to address minority housing needs through large-scale public housing projects. But by the late 1980s, these projects were decaying and stagnant—a living symbol of the failures of the liberal welfare state. Therefore a new approach was needed. "No longer would public housing be at the top of the liberal Democratic agenda," Makin explained. "Instead, borrowing from conservative ideas about the inestimable benefit of home ownership to the striving poor, the Clinton administration and members of his party in the House and Senate decided to use government power to achieve that aim."[5] In a word, Clinton's new housing agenda, like his later embrace of welfare reform, was an example of "triangulation."

Clinton had brought with him to Washington a corps of committed liberals who shared his view that access to affordable housing—which meant access to government-backed mortgage credit—was (like health care and abortion on demand) a civil right. They were determined to use any means necessary—litigation, threats, new regulations, the full power of the federal government—to push their great experiment in social engineering. All, as we shall see, were thoroughly Clintonian, which is to say that their high-minded ideals were sometimes at odds with a tendency to cut ethical corners in their personal and professional lives.

Among the most important of these was Clinton's attorney general, Janet Reno, who would serve as the blunt instrument enforcing Clinton's efforts to push the banking industry into line with the minority housing agenda.

At first glance, Reno might have seemed an unlikely ally in this cause. During her tenure as a prosecutor in south Florida, she had had a rocky relationship with the African-American community. In 1979, when five plainclothes Miami-Dade police officers mistakenly raided a house they believed held a drug cache, the two men inside resisted and were badly beaten. The case immediately took on racial overtones; the police officers were white, and the victims were black. The men argued that they had resisted because they believed the plainclothes officers were robbers. The police had beaten them rather severely. Reno looked into the case and decided she couldn't prosecute the officers. "They made a mistake," she said. "They went to the wrong house by mistake. There was no criminal intent." The black community erupted in anger.

In another instance, Reno had charged several respected black leaders in Miami with corruption and theft. A county commissioner was put behind bars for running illegal bingo games, and the school superintendent was charged with diverting public funds to construct a weekend home. Due to a feeling that she was unfairly singling out blacks, the name "Reno" became an epithet in parts of the city. Garth Reeves, the black publisher of the *Miami Times*, remarked that Reno was to the black community "what Hitler was to the Jews."[6]

But now, as U.S. attorney general, Reno was ready to take on racism in banking. She had reportedly been Hillary Clinton's pick for attorney general, and if nothing else she brought a certain zealotry to fulfilling her commitments. "Credit is the lifeblood of this nation," she said. "We are here to say if voluntary compliance is not forthcoming, we are prepared to take whatever enforcement action is right and proper under the law."

Reno was aided in her efforts by Deval Patrick, who would go on some years later to become the governor of Massachusetts. Patrick became the new civil rights chief in the Clinton Justice Department and carried out Clinton's plans with a deep sense of mission. At his swearing in he promised no less than to "reclaim the American conscience." A former lawyer for the NAACP, Patrick had grown up poor in Chicago and claimed to see rampant racism all around him. "To understand civil rights, you must understand how it feels. How it feels to be hounded by uncertainty and fear about whether you will be fairly treated. How it feels to be trapped in someone else's stereotype, to have people look right through you."[7]

Americans believed in "equality, opportunity, and fair play," Patrick said. "And yet at the same instant we see the racism and unfairness around us. . . . Our mission is to restore the great moral imperative that civil rights is finally all about." He mentioned banks in particular, noting that he felt "a personal commitment" to fair lending.

Patrick also had a personal commitment to his own financial well-being and seemed particularly adept at merging the two.

While at Justice, he handled cases against Texaco, Coca-Cola, and the subprime lender Ameriquest on the ground that they engaged in racial discrimination. Shortly after leaving Justice, he would mysteriously be hired by all three to help them deal with their civil rights problems.

In 1998, while monitoring Texaco's employment practices as part of a court-approved discrimination task force, Patrick took a call from Texaco CEO Peter Bijur, who offered him a job as general counsel for the company. One year later Patrick left the civil rights division and moved into the corporate suite, where he helped the company handle the legal fallout from the discrimination case. Next he went to Coca-Cola, another company that had been scrutinized while he was in the civil rights division. He also helped it handle issues related to racial discrimination and hiring. Most glaringly, perhaps, while at Justice he had sued Long Beach Mortgage Company on grounds of racial discrimination in its lending practices. Like many others, the company settled for $3 million rather than fight the federal government. Long Beach Mortgage became Ameriquest, and a few years later Patrick joined the board of the parent company, for which he was paid $360,000 per year.[8]

When he ran for governor of Massachusetts in 2006, Patrick resigned from the board. A year later, when Ameriquest was in desperate need of cash because of the subprime crisis, the new governor picked up the phone and called Robert Rubin, who was now a senior adviser at Citigroup. Ameriquest had approached Citigroup for an infusion of cash. Patrick vouched to Rubin for "current management and the character of the company."[9]

Clinton installed other powerful allies at the Department of Housing and Urban Development. The first secretary of HUD was Henry Cisneros, the former mayor of San Antonio, one of the highest-profile Hispanics in the country before his career imploded in scandal. Cisneros had an easy charm about him and, most important, the ear of President Clinton who was a personal friend.

His top deputy on housing rights was Assistant Secretary Roberta Achtenberg. An activist lawyer and former San Francisco supervisor, Achtenberg had made headlines by championing efforts to force the Boy Scouts of America to accept gay scoutmasters. As cochair of United Way's gay-issues task force, she promised that when confronting the scouts she would be "holding the money in the left hand and wagging the finger with the right."

Achtenberg, an aggressive political activist, proclaimed that she had developed her social conscience very early. She recalls as a seven-year-old refusing to enter a restaurant because there were picket lines outside. She studied at Berkeley and worked as a legal intern on welfare rights cases. She had gone to law school "because I thought it was a way of doing social justice."

Achtenberg had an expansive view of civil rights. When San Francisco public health director Mervyn Silverman ordered the city's bathhouses closed due to the rapid spread of AIDS, Achtenberg fought the action, declaring that the bathhouses were "institutions of tremendous symbolic significance to a sexual minority." It was a civil rights issue, she said. Achtenberg was shrewd and effective in the way she fought her opponents. When she ran for a seat in the California State Assembly against John Burton, an establishment Democrat, she explained quite publicly and loudly on several occasions that it would be "inappropriate" to bring up Burton's drug and alcohol problems during the campaign.[10]

Early on in the Clinton administration, Achtenberg laid out her plans at a HUD-sponsored summit in Cleveland. The federally funded event was organized by the National Association of Human Rights Workers, the National Fair Housing Alliance, and the International Association of Official Human Rights Agencies. Benjamin Chavis of the NAACP told those assembled that housing was a "life-and-death issue." Then Achtenberg stepped to the podium and explained the Clinton administration's view that housing was a civil right and deserved the same federal guarantees as other fundamental rights. "Housing is a make-or-break, quality-of-life issue, the right to choose where we live is as important as the right to equal educational and employment opportunity and the right to vote," she said. "Yet fair housing historically has been the last civil right to be recognized and the most difficult to secure. . . . Today signals a new stage in the battle for fair housing. The federal government is joining you—to carry its share of the load."[11]

To justify their jihad against banks and other lenders, the Clinton administration embraced a new study by the Boston Federal Reserve

Bank that was then receiving considerable press attention. This "extraordinary 29-page document," as Steven Malanga reported, "codified the new lending wisdom." The study had examined 6 million 1991 home mortgage applications and found that at constant levels of income, 30 percent of blacks and 27 percent of Hispanics were refused loans compared with only 17 percent of whites. When criteria such as credit history, net worth, age, and education were factored in, blacks were still rejected 17 percent of the time compared with 11 percent for whites. How to explain this relatively small 6 percent difference? Immediately it was assumed that this was racism at work. Conventional lending criteria were objectively discriminatory because they did not take into account "the economic culture of urban, lower-income and nontraditional customers." This produced an unconscious bias in lending patterns that was tantamount to racism. "No more studies are needed," said Richard Syron, the president of the Boston Fed. The media hyped the claims: "There's No Whites Only Sign, But . . ." read a headline in *BusinessWeek*.[12]

Meanwhile, the lead author of the study, Alicia Munnell, would join the Clinton administration as assistant secretary of the Treasury.

The Clinton administration cited the data to justify its coming attack on lending standards. "Home ownership is one of our cherished American dreams," said Janet Reno. "These statistics are of great concern to us." The results, she said, spoke for themselves. Kweisi Mfume, a former congressman who was now president of the NAACP, declared, "This data clearly indicates that the American dream of homeownership is a nightmare for people of color."

Henry Cisneros cited the data publicly on several occasions, assuming, like everyone else, that the gap in lending was the obvious result of racial discrimination. It could not be explained in any other way. "It's time to put an end to discrimination in mortgage lending," he declared. "We can do better. We mean business by getting tougher with mortgage companies through HUD's Mortgage Review Board." [13]

But once again, the data didn't really support the conclusion that racism was at fault. There were also some glaring statistical errors in the study. David Horne, an economist at the FDIC, examined the data and found that forty-nine of the seventy banks had not rejected any minority loan applicants and that two of the remaining twenty-one were responsible for more than half of the denials to black applicants. One of the two was a minority-owned bank. The other had an extensive minority outreach program. Horne also found that many of those who were denied loans had negative net worth and ought to have been excluded from the study, and that often applicants were denied because of the lenders' inability to verify the information that they had been given by the applicant.

Professor Harold Black took the same Boston Fed model and conducted another study. Intriguingly, he found that black applicants were denied loans at significantly higher rates by black-owned banks than by white-owned banks.[14] Presumably the managements of those banks felt less vulnerable to racial blackmail and were therefore more prone to rely on rational lending criteria.

Two reporters for *Forbes* also looked at the data and found that blacks and whites who were granted loans had the same default rates. If it was harder for blacks to get loans than whites because of racial discrimination, one would expect blacks to have a lower default rate on average. The fact that both had the same rates showed that the market was working. The Nobel laureate Gary Becker, who essentially pioneered the study of discrimination economics, stated that the study had enormous problems. "The flaw in all studies of discrimination by banks in applications for mortgages is that they have not determined the profitability of loans to different groups. . . . A valid study of discrimination would calculate default rates, late payments, interest rates and other determinants of the profitability of loans." [15]

Those who raised such objections were browbeaten by members of Congress. When Federal Reserve Board Governor Brock LaWare said that policy makers should not be too quick to jump to conclusions about widespread discrimination, Senator Carol Moseley Braun, a Democrat from Illinois, reprimanded him in her usual tone of high dudgeon: "This is not only unacceptable, but this kind of ignorance coming out of our regulatory agencies is just not to be tolerated."[16]

Regardless, the battle was joined. Reno and Cisneros went to Capitol Hill and, shoulder to shoulder in front of the Senate Banking Committee, laid out their plans. "The message on fair lending from the Clinton administration is simple and clear," said Cisneros. "We are changing the way we do business, and we mean business." Reno said that she expected banks and lenders to offer more loans—or else. "While it can be expensive . . . the power of persuasion in the courtroom cannot be denied," she told the Senate. "I am confident we will produce unprecedented results."[17] Committee members including Senators Donald Riegle, Paul Sarbanes, Alfonse D'Amato, and Patty Murray listened and applauded what they heard.

Of course, the fact that Reno embraced the use of statistics to prove discrimination is filled with irony. While a prosecutor in Florida, she had been criticized by the U.S. Commission on Civil Rights for prosecuting three times as many blacks in juvenile court as any other racial or ethnic group. The data had been offered as proof that her office was engaging in discriminatory activities. At the time, Reno had criticized the report, saying that the findings were misleading.[18]

Reno immediately hired eighteen more lawyers in the housing section of Deval Patrick's Civil Rights Division and sent them off after banks and lenders. The Justice Department drew up a list of two hundred lenders considered "ripe for investigation." They instructed bank regulators to look into them and asked for the names of four or five that they could sue to make an example of.[19]

Many of the cases were handled by Paul Hancock, a longtime Justice Department lawyer who headed the Fair Lending Enforcement Program. (He would later, as a private attorney, help to argue *Gore v. Bush* before the Florida and U.S. Supreme Courts.) Hancock made it clear that anyone could be a target. "Small banks shouldn't feel they have a safe harbor," he said. "We're not eliminating anybody from the mix here. Lending discrimination will be challenged regardless of where it is occurring."

On November 16, 1993, the Justice Department filed charges under the Equal Credit Opportunity and Fair Housing Acts against tiny ($18 million in assets) Blackpipe State Bank of Martin, South Dakota. Why pick on Blackpipe? There had been criticisms leveled against the bank in American Indian publications, which claimed it was reluctant to provide mortgage loans on a nearby reservation. The Clinton Justice Department agreed and filed suit.

In its defense, the bank maintained that lending on reservations was fraught with confusion and difficulty. There were jurisdictional problems; the status of tribal courts was quite different than that of state or federal courts. There were also unique land issues. Who might own property on reservations was often open to question, determined by old treaties, federal laws, and the descendents of the original landowners. Much Indian-owned land was held in trust by the federal government. Indeed, as bank examiners had explained in a 1990 report regarding Blackpipe, "The bank no longer makes these types of loans [mortgages] because of legal problems encountered pertaining to the foreclosure process when the property involves Native American trust land. . . . Bank management has attempted to resolve this legal impediment with the Bureau of Indian Affairs, but no remedy was achieved. The bank refers all consumer real estate mortgage inquiries to federal agencies that would be able to meet these credit needs." [20]

The Clinton administration filed suit anyway.

The Justice Department also went after First National Bank of

Vicksburg, Mississippi, on the ground that it was overcharging black customers for home improvement loans. Another lawsuit followed against Chevy Chase Federal Savings Bank, the largest lender in Washington, D.C. Because Chevy Chase made 97 percent of its loans in predominantly white areas, it was engaging in "redlining," according to the Justice Department. All the banks denied the charges, but all of them settled, citing the enormous cost of doing legal battle with the federal government. As the *Miami Herald* noted, "No lender accused of loan discrimination has chosen to fight the Justice Department. It could take millions of dollars in legal fees to fight the government."[21]

Federal officials also launched investigations of Barnett Banks of Jacksonville, Florida, and Shawmut National Corporation of Hartford, Connecticut. Shawmut was eventually denied its application to acquire the New Dartmouth Bank on the grounds that it had a bad "record on racial fairness," as the *New York Times* explained. Amazingly, *there were no individual complaints from minorities about the bank*. No matter: the case proceeded anyway. The threat of similar lawsuits prompted other banks, such as Albank of Albany, New York, to begin offering discounted loans in poor neighborhoods. Janet Reno declared with satisfaction that the case would make home ownership "more likely for many more Americans."[22]

As a result of this legal offensive, the banking industry began to lawyer up. The Savings & Community Bankers of America formed a "war-chest for fair-lending litigation." The American Bankers Association engaged a major law firm to look into the issue.[23] As the Federal Reserve Bank reported, "every suit brought by the U.S. Justice Department has been the focus of great interest, at least among lenders."[24]

Activists had been using the CRA to great effect, but now the Clinton administration was moving on a whole new front. In a kind of tag team effort, Reno worked closely with the National Community

Reinvestment Coalition (NCRC), which bills itself as "the nation's largest economic justice trade association," comprising six hundred member organizations "dedicated to increasing access to capital and credit for working class and minority communities." NCRC would target a bank, and Reno would start turning the wheels of litigation. Meanwhile, Housing Secretary Cisneros met monthly with representatives from ACORN.

The CRA applied only to chartered banks; mortgage companies were exempt. But the Clinton legal team was going after everyone— banks, mortgage companies, anyone that was making loans. And as they using the Fair Housing and the Fair Credit Acts as the legal bases for their actions, everyone was in their crosshairs.

Federal banking authorities also began using undercover agents to find out if mortgage lenders were discriminating.[25] How did they determine discrimination? One of the more troubling aspects of the Clinton legal offensive was how they began to change the meaning of this term. The traditional view was that policies designed explicitly to turn away minorities, or attitudes that precluded lending to minorities, were considered to be discriminatory. The Clinton administration was now changing the definition in a manner that made it possible for bankers to be guilty of discrimination without intent— indeed, without even knowing that they were discriminating. The administration called it "disparate impact."

According to this theory, a bank might have policies that were fair and equal, but if the outcomes were different, the policy was "objectively" discriminatory. For example, if a lender had a long-standing policy of not issuing mortgages on homes worth less than $60,000, it could be guilty of discrimination because that policy would "disproportionately and adversely" affect the poor and certain minority groups. This was also known as "institutional racism." A lender could apply the same policy to all borrowers, but if one group felt the impact more than others, it could be guilty of discrimination.

And according to the working definition established by Janet Reno, it would be up to the banks to prove that they were not engaging in discriminatory behavior—not the other way around.[26]

Federal regulations now declared that the government would use "evidence of 'disparate treatment' to determine whether the Fair Housing Act was being violated." "A clear definition of lending discrimination is vital," offered Henry Cisneros. "As of today, we have it." And it became the guiding principle of the Clinton administration.[27]

The Clinton push for numerical guidelines "will result in quotas," Kevin Kane, the president of CRA Consultants, told *Mortgage Banking.* "[Banks] will throw away credit criteria." If told that they don't have enough minority loans, "the banks will just do it, make 20 loans to minorities and consider them hazardous or throwaway loans." Lawrence B. Lindsey, then a governor of the Federal Reserve Board, warned, "A total reliance on statistics in credit enforcement will ultimately lead to a complete replacement of bank judgment and reason regarding loan approval with statistical rules."[28] It was, wrote Robert Stowe England in *Mortgage Banking*, another example that "American society is making major decisions based not on character but on finely tuned racial statistics."[29]

The Justice Department got plenty of help from a host of federal regulatory agencies that traditionally had performed fairly mundane functions. Eugene Ludwig, Bill Clinton's comptroller of the currency (who regulates national banks), borrowed from the activists' dictionary when he spoke about "the democratization of credit."[30] In May 1993, he began using undercover testers "to expose biased lending." The Office of the Comptroller of the Currency reported that it was "working closely" with members of Congress, such as Representative Maxine Waters—a leading member of the Congressional Black Caucus and one of the key voices supporting the antiredlining campaign—on how to best conduct the tests. Astonishingly, in the end it was decided that any bank with just five minority denials in mortgage loan applications would face a fair lending exam by one of the federal government's 1,900 bank examiners.[31] The Federal Department Insurance Corporation (FDIC) launched a sweeping probe into lending bias, showing up unannounced at several banks and subjecting them to "in-depth investigation, which could result in a formal enforcement action or referral to the Justice Department." The National Credit Union Agency likewise set up its own lending enforcement program.[32]

By 1995, Reno could boast that the Clinton administration "has brought hundreds of cases" into the courts on behalf of minorities who had been "rejected for a loan because of their race, gender or national origin."[33] Ambitious state attorneys general around the country also got into the fray. Pennsylvania's attorney general investigated several lenders and promised more lawsuits.[34]

On Martin Luther King Day 1994, Clinton signed an executive order creating a presidential Fair Housing Council, to be headed by HUD Secretary Henry Cisneros. The order was signed with the promise "that for the very first time puts the full weight of the federal government behind efforts to guarantee fair housing for everyone." Said Roberta Achtenberg, the assistant HUD secretary for fair housing and equal opportunity, "Today signals a new stage in the battle for fair housing." Achtenberg formed a special unit to sniff out discrimination in mortgage lending and promised that the "first job will be to promulgate substantive regulations" in the area.[35]

Cisernos was tasked with signing seventy-five voluntary agreements from private lenders to comply with federal fair lending laws. However, "voluntary" was not quite how many in the industry saw it. As the *ABA Banking Journal* noted, "Bank lenders involved in the negotiations complained that HUD representatives were trying to ram the 'voluntary' agreements down their throats."[36]

By early 1994, the Mortgage Bankers Association of America had signed a three-year master best-practices agreement with Cisneros. Members were essentially given the option of signing the agreement or risking legal action. The first to sign was the mortgage giant Countrywide.

Assistant Secretary Achtenberg set up ten enforcement centers around the country to sniff out alleged financial racism. HUD actively collaborated with the NAACP in bringing class-action lawsuits against mortgage brokers.[37] It actively encouraged people to file discrimination complaints against lenders, and predictably, such complaints surged by 46 percent in the first year of the Clinton administration.[38] In March 1994 the administration laid out its criteria for determining whether a lender was discriminating. How far was HUD prepared to go? It got to the point where it interpreted the Fair Housing Act as saying there needed to be quotas for African-American models in all housing display ads. The *Washington Post* complied to avoid litigation, agreeing with a local activist group that the newspaper would have a minimum of 25 percent African-American models in all ads.[39]

The benchmark for the Clinton administration was the National Homeownership Strategy, also spearheaded by Cisneros. The goal was to dramatically increase home-ownership rates, particularly among the poor and minorities. The only way to do this, of course, was to aggressively loosen mortgage lending standards. The strategy called for "financing strategies, fueled by the creativity and resources of the private and public sectors, to help homeowners that lack cash to buy a home or to make the payments." It seems never to have occurred to them that people who couldn't make payments . . . *shouldn't own homes.*[40]

"We were trying to be creative," Cisneros would say later. The Clinton administration was now walking the same path as the activists: reduce and loosen credit standards; let the poor assume a greater debt burden; don't worry about the consequences. Federally backed loans through HUD would no longer require that a borrower prove five years of stable income. Now only three years would do. Mortgages used to require looking at the income of a husband and wife. But now banks were "willing to count the earnings of all the siblings, in-laws and adult children" who happened to be living together. This was done in recognition of the fact that "some Hispanics live in extended families." First Union, one of the country's largest lenders, fearful that it might face legal action by the federal government, told loan officers to examine every loan and to "look at every possible way to make that loan." Cisneros also allowed lenders to hire their own appraisers rather than rely on those selected by HUD. The move was designed to save low-income buyers money. The trouble was that it created opportunities to inflate appraisals.[41]

The Clinton administration aggressively pursued this path even as its own data started to indicate that the "affordable mortgage loans" provided through the CRA had higher-than-normal default rates. "This delinquency rate increase must be taken seriously," noted the comptroller of the currency, Eugene Ludwig—before advocating for even more such risky loans.[42]

After Cisneros left HUD in 1997, he accepted a lucrative appointment on the board of the subprime lender Countrywide, a prominent player that will resurface in our story. He was replaced at HUD by Andrew Cuomo, the son of former New York Governor Mario Cuomo, a young man in a hurry with considerable ambitions of his own.

Cuomo quickly picked up where Cisneros had left off. The difference was that he lacked Cisneros's calm style and proceeded with a much more brazen attitude. Thus on April 6, 1998, he held a remarkable press conference at which he announced a settlement with a major bank in a lending discrimination case; Cuomo explained that the bank would now be offering $2.1 billion in new subprime loans. Then came a series of stunning admissions: those receiving the new loans would not have qualified "but for the affirmative action on the part of the bank"; the loans were at "higher risk of default"; and, yes, the bank was going to have to "lower its standards on loan qualifications" to make the loans.

In his missionary zeal, Cuomo saw all of this as a good thing. He would settle numerous other cases against other financial institutions in a similar fashion. Like so many involved in these shakedowns, Cuomo knew what he was doing. That is to say, he knew perfectly well that this vast social experiment would cost the banks enormous sums of money. But he didn't seem to care. To the contrary, he seemed to think they had it coming.

Cuomo also pushed the Federal Housing Administration (FHA) into supporting no-money-down, no-closing-cost mortgages while legalizing fees to brokers who pushed the mortgages (a practice one federal judge has called "kickbacks"). Brokers now had powerful financial incentives to push these toxic products.[43]

To meet the growing demand for loans to the poor and minorities, many of whom had poor credit, bankers and lenders embraced subprime lending, a risky innovation that was being pushed by the housing activists. Subprime loans are simply loans given to those with a low credit score or a bad credit history. The loans often come with a higher interest rate or higher fees because the risk of default is higher.

It was already well known that subprime mortgages had high failure rates. In Baltimore, for example, subprime loans accounted for 21 percent of all mortgages but 45 percent of the foreclosure petitions in the city. Still, activists and their enablers in the government proceeded to push such loans in the name of financial democracy.

Needless to say, these aggressive low-income and minority home-ownership goals generated concern and anxiety in the business community. According to a HUD report, "the increased goals created tensions in its business practices between meeting the goals and conducting responsible lending practices."

Private-sector economists were worried, too. Cynthia Latta, an economist with DRI/McGraw-Hill, said in 1999, "Banks are under a great deal of pressure to lend in these communities. It is very political. But I still have reservations about whether you're really doing anyone a favor by letting them borrow 100 percent of the cost of a home. It makes it so easy for them to get in over their heads. . . . We have created a tremendous amount of risk. At some point the economy is going to turn down. There will be large numbers of defaults that will trigger a lot of political heat."[44]

By early 1995, the Clinton administration was pushing a revision to the CRA. It wanted a numerical quota system to be used to evaluate whether banks were loaning enough to low-income and minority groups. Up to this point, federal regulators had interpreted the CRA in an ambiguous manner. The law was somewhat vague, and whether a given bank was lending enough to these groups was often in the eye of the beholder. Now the Clinton administration had decided that banks needed to prove that they were making loans in proportion to the demographics of their area of operation. The old method of looking at a bank's portfolio in general was changed to measuring "diversity" based on raw statistics. "If a bank has, for example, a 15% share of the market, it should have roughly a comparable share in poor neighborhoods," it was explained.[45]

But now there was resistance on Capitol Hill. Republicans had taken over Congress in 1994 and were actually talking about abolishing the CRA.

Conservatives such as Senator Phil Gramm of Texas and Congressman Bill McCollum of Florida pushed for legislation to repeal the act or at the very least dramatically scale it back. On the House floor McCollum argued that complying with the CRA was costing banks $1 billion a year and that with the new Clinton standards this truly was "credit allocation, pure and simple." He charged that Clinton had turned the Justice Department into a "bank regulator."

Predictably, Democrat Congressmen Barney Frank and Henry Gonzalez vigorously defended the CRA. Republican efforts to dismantle the CRA were "simply shameful," said Gonzalez.[46]

Activist groups such as ACORN quickly mobilized. When a congressional committee held discussions about defanging the CRA, ACORN national president Maude Hurd and a large group of activists barged in and began chanting "CRA has got to stay!" and "Banks for greed, not for need!" They tried to grab the microphone, but Hurd and five other ACORN activists were arrested. The U.S. Capitol Police put them into jail for trespassing and causing a public disruption. Senator Ted Kennedy and Congressman Joe Kennedy tried to get them released but failed. Only when Congressman Maxine Waters showed up and threatened to stay until they were released were the protesters let go.[47]

In the end, instead of pulling the CRA's teeth, the GOP-dominated Congress turned a blind eye to the Clinton administration's efforts to strengthen it. Along with a new numerical quota system to determine whether someone was complying with the law, Clinton did something else: he mandated that CRA ratings be made public. No longer would activist groups have to shoot in the dark with allegations that banks were not doing their share. Now they could look at government evaluations directly. This in turn fueled greater activism. As the industry publication *Mortgage Banking* noted, "Pressure from community groups for greater CRA activity has increased over the years, and banks have been encouraged to explore innovative ways of meeting their communities' credit needs and providing more loans to minority borrowers." Those creative ways were, of course, the softening of lending standards.[48]

The provisions put enormous power into the hands of community organizers, who could seriously damage banks and lenders they believed were "not doing enough" for the poor. A good CRA rating was necessary if a bank wanted to have a merger approved, expand, or even open new branches. It might also face direct legal challenges by the Justice Department. This allowed community activists to intervene at yearly bank reviews, questioning banks' lending practices and accusing them of racism.

Needless to say, banks felt the pressure even more and began giving in to the most extreme demands of the activists. In 1995, Wells Fargo agreed to a $45 *billion* arrangement with the California Reinvestment Committee. Washington Mutual agreed to a $120 *billion* agreement with activists in Washington State and California. In Ohio, Star Bank signed a $5 billion agreement with the Coalition on Homelessness to make loans available to the homeless in "targeted areas." (Think about it: loans to the homeless.) U.S. Bank agreed to open an "alternative loan division" for those who had poor credit and couldn't qualify for an ordinary mortgage. The subprime lender Household International agreed to provide $3 billion in subprime loans to those with "blemished credit histories" and further increased its risk by offering these individuals lower interest rates.

Soon banks abandoned almost completely any notion that borrowers should put much money down. In 1999, Sovereign Bank agreed to offer 95 percent Loan-to-Value mortgages to borrowers with poor credit history and to provide "up to 3% of the down payment in the form of a gift or a second mortgage." In other words, a borrower would have hardly any skin in the game. It used to be that banks did not want a mortgage payment to account for more than 25 to 28 percent of a person's monthly budget. But activists believed that this was discriminatory to the poor. So in 1998, the banking giant First Union was harassed by activists and forced to offer "affordable home loans with no private mortgage insurance, down payments coming from sweat equity, debt-to-income ratios as high as 33 to 38%." The bank also agreed to provide "100 percent Loan-to-Value product with no down payment, no points, no private mortgage insurance, and $500 of the borrower's own funds at closing which can be a gift or paid by the seller." Other banks were quick to follow suit.[49]

Everyone involved knew that these loans were fraught with peril. These CRA loans usually had a high percentage with "less than a 660 FICO score," denoting poor credit, explained Dale Westhoff, a senior managing director at Bear, Stearns. Such loans also had "a disproportionately high number" with very high LTVs, a key determinant in default.[50]

Heedless of risk, activists and Clinton administration officials were eager for the mortgage spigot to remain open. So they pushed financial institutions into what they called "secondary market" agreements: banks could sign mortgage agreements and hold on to them, tying up their money for years, or they could go to the secondary market and sell their mortgages through so-called mortgage-backed securities. The National Community Reinvestment Coalition advised its members: "CRA agreements should, where possible, include commitments by secondary market institutions to purchase loans so that the banks can obtain more capital for making additional CRA loans."

In other words, it was pushing for banks to create mortgage-backed securities based on subprime mortgages, so that the capital could be used for more subprime mortgages.[51] As we now know, the ultimate consequences of this fateful move were catastrophic.

Ellen Seidman, who served as special assistant for economic policy to President Clinton, would later declare that this had been a major accomplishment of the Clinton administration. She touted the fact that the administration had encouraged and pushed for the greatest use of mortgage-backed securities: "Fannie Mae, Freddie Mac, investment bankers and mortgage companies are offering targeted, mortgage-backed securities that enhance liquidity and increase the capital available for community development. Investors are finding that these securities—particularly the targeted, mortgage-backed securities—have very attractive characteristics." Then she added this about the Community Reinvestment Act: "Yet, without CRA as an impetus, this market would likely not have developed." She explained with apparent excitement that "mortgage bankers have responded to the improved security market opportunities CRA has fostered. The market for whole loans and mortgage-backed securities targeted to lower-income borrowers is booming, and mortgage companies are aggressively pursuing this business."[52]

A few years later, during the economic meltdown of 2009, these mortgage-backed securities would be denounced by indignant Democrats as the product of unfettered greed by Wall Street speculators. Wall Street was hardly innocent in the matter. But the leading role of Democrats and their allies in the fair housing movement in vigorously pushing these risky products on the banking industry has gone strangely unremarked.

Bankers themselves saw the Clinton administration agenda for what it was: a blatant attempt to bankroll Clinton's social agenda on the back of the banking sector. "They seem to be trying to push the financial industry into making more risky loans and accepting more risky loans than they're willing to take," complained John Shivers, the president of the Independent Bankers Association of America. "With the federal budget constraints, there's not enough money to pay for the social programs the Clinton administration wants. They want the financial services industry to pay for them."[53]

Little surprise that mortgages to the poor and minority communities skyrocketed. A study by Harvard University found that during the Clinton years, home loans to the poor, particularly subprime lending, "increased dramatically" as a result of Clinton policies. The National Community Reinvestment Coalition boasted that $1.09 trillion in CRA loans had been negotiated during Clinton's first few years.[54] By the end of the Clinton administration the raw number of dollars committed under the CRA was rising dramatically:

$221 billion in 1997, $812 billion in 1998, another $103 billion in 1999.

The results were exactly what the activists wanted: the number of minority home owners soared. Home loans to blacks grew by 72 percent and to Latinos by 45 percent. The number of blacks owning their own homes increased three times as fast as that of whites; the number of Latino home owners grew at five times the rate of whites. Mortgages in low-income neighborhoods increased by 40 percent during the first two years of the Clinton administration. For families earning less than the median income, home ownership rates increased at more than three times the rate of other Americans. But much of this was happening because of the explosion in subprime lending. According to Freddie Mac, 40 percent of black Americans had credit delinquencies, a bankruptcy filing, or a history of late payments. Studies showed that blacks had more debt, less savings, and fewer assets than most Americans. Yet they were being actively encouraged to take on large mortgages.[55]

Much of this was accomplished by pushing subprime loans on the reluctant banking sector. Subprime lending exploded during the Clinton years. It increased *twentyfold* between 1993 and 2000, and the number of subprime lenders surged from a handful to more than fifty.[56]

The great Clinton social experiment was wildly ambitious and was carried out with typical grandiosity in the name of civil rights and social justice. In remarks before the National Community Reinvestment Coalition, Janet Reno explained, "I want to thank you for your work in economic justice . . . day in and day out, you have been on the front lines in pressing for the kinds of banking services that potential homebuyers need. . . . And I just want you to know how much I appreciate it." Reno described the CRA itself not as a legislative cudgel to be wielded against banks but as a helpful tool for lenders. "The new Community Reinvestment Act regulations enable

lenders to develop customized strategic plans for meeting their obligations under the act, and many have been developed in partnership
with your local organizations."

Somehow she managed to see this coercive abuse of government
power as a positive and helpful partnership, beneficial to all parties.
To paper over the yawning cognitive dissonance in her description of
reality, Reno deployed the new vocabulary of enlightened capitalism.
Thus she explained that they were getting results with banks because
"most bankers want to be good and responsible corporate citizens,
or they're willing to be if they're nudged in the right direction . . .
by Justice Department lawyers who care and want to do the right
thing."

Responsibility. Corporate citizenship. Doing the right thing. These
coinages—along with the idea that it is government's role to "nudge"
private actors in "the right direction"—were the unique contribution
of a generation of liberal statists intent on using the power of government to meet progressive goals. What's more, they were amazingly
successful. Reno noted with genuine pride that in the first four years
of the Clinton administration, there was an 86 percent increase of all
bank commitments under the act and that,under her direction, DOJ
had filed thirteen major fair lending lawsuits. "We will continue to
focus on discrimination in underwriting, the process of evaluating
the qualifications of credit applicants." [57]

By the end of the Clinton administration, Treasury Secretary
Robert Rubin could proudly announce that "the number of home
mortgage loans extended to African-Americans has increased by almost 60 percent, to Hispanics by a little over 60 percent, to low and
moderate income borrowers by a touch under 40 percent, figures
that are well above overall market increases." He regarded the Community Reinvestment Act as "one of our most effective tools" in this
effort.[58]

Rubin praised the CRA, declaring that it "had greatly increased the availability and flow of credit in inner cities."[59] In Congress, supporters of the act lauded it as a program that advanced a public agenda using private dollars. Congressman Henry Gonzalez, a Democrat from Texas and, Chairman of the House Banking Committee, explained during a debate on reforming the CRA that "it is critical that we save a tried-and-true program that relies on private dollars."[60] None of these comments acknowledged that this glorious new era of cooperation had been brought about by hardball tactics of coercion.

After he left the Clinton administration and went to work at Citigroup, Rubin continued to burnish his progressive credentials as the chair of the Local Initiatives Support Corporation, which finances affordable housing and urban renewal projects and advocates on behalf of the CRA.[61]

This massive experiment in socially engineered housing equality created a whole new class of debtors in America. And by far the greatest victims were the very people Clinton was trying to help. Democratizing finance really meant "democratizing" credit, by giving subprime loans to poor people with bad credit. While the ratio of mortgage debt to overall debt declined for the middle class and wealthy during the Clinton years, it exploded for the poor. In less than a decade, the poor had mortgage-to-debt ratios substantially higher than that of other economic group. As Raisa Bachieva, the research director at the New York City Department of Housing Preservation and Development, explained, "The federal government, through the Community Reinvestment Act and the affordable housing mandates for Fannie Mae and Freddie Mac, has substantially increased the incentives for regulated banks and secondary mortgage market agencies to make loans to low- and moderate-income people and people living in low- and moderate-income neighborhoods. . . . A large part of this outreach was undoubtedly attributable to the efforts of the Clinton administration to enforce the Community Reinvestment Act of 1977."[62]

One might think that all this debt was a good thing; after all, the poor were investing in a home, creating incentives to seek and hold steady employment, thus putting their feet on the ladder to economic advancement. But a lot of research shows that this is actually bad for the poor. As Raisa Bachieva argues, the poor are highly vulnerable to even small income shocks that push them into bankruptcy. If they are renting, they have the flexibility to deal with the challenge. But if they are saddled with a mortgage, it can present serious problems. This is particularly the case when social engineers are encouraging the poor to stretch, assuming mortgages and other ownership costs they normally couldn't afford.[63]

Bachieva also points out that many of the poor took out mortgages with high loan-to-value ratios; in other words, they put very little—if anything—down on their mortgage. This is exactly what the activists and their Clinton administration allies were actively encouraging. But loan-to-value ratios are a key determinant in knowing when homeowners are at risk of default. Those with high LTV ratios "have a significantly higher incidence of default and foreclosure." Because borrowers have very little of their own money at risk, they can easily walk away.[64]

A handy guide to the roots of the impending collapse may be found in the official statements of Ellen Seidman, who had served as special assistant for economic policy to President Clinton and was later head of the Office of Thrift Supervision. We have already seen how she wanted to claim credit for the explosion in mortgage-backed securities. But it is interesting to note that she claimed credit at the time for each component that contributed to the looming subprime crisis, seeing them each in turn as major breakthroughs.

Thus she announced approvingly that "homeownership is at an

all-time high." She noted with pride that it was the Clinton administration that had loosened lending standards. "Financial institutions have responded by revising their underwriting practices, making lending standards more flexible," she explained in an article in *Mortgage Banking*. "Smaller down-payment requirements, more liberal rules on contributions to closing costs and reductions in cash reserve requirements have helped to lower the 'wealth hurdle' to homeownership." She noted that banks that were subject to the CRA were devoting a greater share of mortgages to low-income borrowers. But the data also "indicates that lenders not covered by CRA have devoted a growing proportion of their home-purchase lending to low- and moderate-income communities and borrowers."[65]

Using the full force and power of the federal government, the Clinton administration propelled America's banks into a dangerous new direction, forcing them to make trillions of dollars in loans to risky customers with poor credit histories. But the activists and the Clinton administration were not yet finished. Next they planned to hijack Fannie Mae and Freddie Mac.

COVER YOUR FANNIE

How Fannie Mae and Freddie Mac Were
Taken Over by Liberal Activists

I want to roll the dice a little bit more in this situation towards subsidized housing.

—CONGRESSMAN BARNEY FRANK, 2003

In 2002, the National Training and Information Center celebrated the life and work of Gale Cincotta with a lavish event at the Chicago Cultural Center. More than two hundred activists, bankers, politicians, and bureaucrats showed up to honor Cincotta, who had passed away just a year earlier.

There was nothing surprising or unusual about this. The activist community founded by Saul Alinsky and his acolytes had grown into a thriving national enterprise with hundreds of local community organizations and umbrella groups all over the country. They had very good reason to honor Cincotta, whose efforts—particularly in pushing the CRA through Congress—had done so much to empower the movement.

What *was* unusual was the sponsor and keynote speaker for this event. Those duties didn't fall to one of Cincotta's old comrades in arms or a prominent radical activist. Instead, they were gladly accepted by the taxpayer-backed financial behemoth Fannie Mae and its chief operating officer, Daniel Mudd.

Mudd, a former director of several large capital funds—and the son of the TV anchor Roger Mudd—gave an emotional speech that was a testament to Cincotta's influence on the direction of the mortgage giant. "Fannie Mae and NTIC were never enemies . . . we just hadn't been properly educated," he remarked in a lame attempt at humor. (The joke no doubt elicited knowing laughter in the room.) According to one published account, Mudd "went on to credit Cincotta with many of the mortgage products and initiatives that help Fannie Mae meet its affordable housing mission today."[1]

This otherwise unremarkable speech in fact marked the complete takeover of the GSEs by the activist wing of the Democratic Party. Fannie Mae had effectively become the property of the Congressional Black and Hispanic Caucuses and their allies inside and outside government.

But that is not the worst of it. The real lesson of the Fannie-Freddie scandal is what happens when government tries to rig the marketplace in favor of a certain socially approved outcome. It is worth paying attention to Fannie and Freddie because their story is about to be replicated by President Obama and his team—which is largely composed of Clinton retreads—in numerous policy areas, including energy, health care, education, and even once again in housing.

Fannie and Freddie used to be steady, even boring entities that simply served to lubricate the mortgage market so that middle-class Americans would find it easier to get a loan. But in the 1990s they were effectively hijacked by liberal activists who helped change their mission, thereby endangering the American financial system. Amazingly, almost nothing has been written about the takeover and transformation of these two powerful semiprivate entities. Furthermore, Fannie Mae and Freddie Mac would go on to play a central role in the rise of notorious subprime lenders such as Countrywide by purchasing billions of dollars in toxic mortgages and fueling their growth.

The Federal National Mortgage Association (Fannie Mae) and the Federal Home Loan Mortgage Corporation (Freddie Mac) are venerable institutions formed during the Great Depression to help stabilize the housing market. Both have long been considered fundamental aspects of Franklin D. Roosevelt's New Deal legacy.

Fannie Mae was established in 1938 to invest in the newly created 30-year self-amortizing mortgage, which was intended to help people keep their homes and stabilize the housing sector. Its mission was funding "mortgage loans insured by . . . the Federal Housing Administration." Fannie Mae's role would be simple. After a bank gave a loan to a consumer, it would sell the loan to Fannie Mae, which would then resell the loan in capital markets around the world. By doing so it was freeing up bank capital to finance more mortgages in the United States. It was also spreading around the risk of the loan, effectively exporting it to the global economic system.

Fannie and Freddie acted liked lubricants of the financial system and helped to make home ownership a reality for millions of Americans. But both were essentially providing a benefit to the American middle class, rather than the poor, because they remained conservative in the mortgages they would buy. Originally, Fannie and Freddie bought only mortgages in which buyers had put a substantial portion of money down and had solid credit histories and a steady income. By 2000, Fannie Mae owned or guaranteed nearly half of all home mortgages in the United States and almost 80 percent of middle-class mortgages.[2]

Like many government agencies, Fannie Mae outlived the crisis it was designed to alleviate. So in 1968 the Johnson administration spun it off as a private company and chartered it as a government-sponsored enterprise (GSE). Two years later, Freddie Mac was like-wise chartered. Fannie Mae sold stock to investors shortly afterward, much of it gobbled up by mortgage companies that already did busi-ness with the company. (Freddie Mac did not become a public com-pany until 1989.)

GSEs are unique hybrids, rare in American life. They are private companies but are implicitly guaranteed by the federal government (that is, by us taxpayers). And that creates a unique set of problems.

Fannie and Freddie are in effect half public and half private. The profits they earn in interest on the mortgages they have purchased or sold to others go to their shareholders, and senior executives receive hefty performance bonuses based on how well the company per-forms. But both have access to a special line of credit worth $2.25 bil-lion from the U.S. Treasury. In short, they can borrow money at much lower rates than any other financial company in America. And up until the George W. Bush administration, they were both exempt from registering and filing with the Securities and Exchange Com-mission. On top of that, they do not have to worry about paying state or local income taxes or meeting the capital requirements imposed on banks. Thus they operate on terms far more advantageous than any other financial entity in America—or the world, for that matter.

But most important, there is a widespread belief that because they are government-sponsored enterprises, the federal government will never let them fail. They will be bailed out if they ever get into serious financial trouble. Therefore they are able to sell bonds with very low yields because investors assume they are extremely safe and backed by the federal government.

Originally, their mission was relatively restricted: to buy mort-gages from banks so the banks could offer more mortgages. More-over, they were run for decades by cautious executives such as David Maxwell and Allan Hunter, former politicians who relied on qualified technocrats and accountants for financial decision making. But in the 1990s, Fannie and Freddie were taken over by ambitious political entrepreneurs and activists who wanted to turn them into financial institutions that would redistribute wealth by taking on the affordable housing mission. As we saw in the previous chapter, the Clinton administration had made affordable housing a major domestic priority. The GSEs were natural allies in this policy and were therefore targeted for takeover by Clinton and his allies on Capitol Hill, particularly in the Black and Hispanic Congressional Caucuses.

By the middle of the Clinton administration, Fannie and Freddie were no longer simple lubricants of the American financial system. Instead, they became ground zero of a vast social engineering project that was willing to take exorbitant risks with taxpayer dollars. Once securely in the hands of idealistic baby boomers who constantly confused their corporate mission with their activist vocation, these institutions soon abandoned their traditional conservative approach to mortgage finance and became avenues for relaxed lending standards, helping in turn to inflate the subprime credit bubble.

The process began with an activist campaign that targeted Fannie Mae with protests and demanded that its mission be changed. Using many of the same techniques that had worked on the banks, groups such as National People's Action and the National Training and Information Center "flooded the corporate headquarters" with letters of protest. They organized protests and demanded meetings with executives. Fannie Mae executives caved in pretty quickly. As National People's Action reported, "From this action, a very successful partnership evolved between Fannie Mae and NPA groups." The activists soon convinced Fannie Mae to start buying mortgages that did not use credit scores as the basis of providing a mortgage. And Fannie Mae started using a mortgage loan product developed by the activists.[3]

Fannie Mae also started to radically lower standards on the mortgages it purchased. The activists helped Fannie Mae set up a program that would "enable individuals and families with low and moderate incomes to purchase a home with as little as $500, or 1 percent of the home purchase price (whichever is less)." The activists and Fannie Mae were now "partners"—in much the same way that the Mafia is a "partner" of the construction and private waste disposal industries.[4]

To celebrate its new role, Fannie Mae began sponsoring conferences organized by National People's Action, at which senior executives and leading political figures would often speak. Senator Chuck Schumer, among others, attended its national conferences and lavishly applauded its efforts.[5]

Meanwhile, organized labor got into the act, pushing hard for the Clinton administration to require Fannie Mae "to target their programs to families and communities with modest incomes and minority areas redlined by major banks."[6]

Environmentalist groups also jumped on board, encouraging low- and middle-income people to buy homes in cities rather than move to the suburbs and cause "creeping sprawl." Fannie was all too happy to partner with radical environmentalist groups such as the Natural Resources Defense Council to "lend more money to people who wish to buy a home in urban areas." Lending increasingly had less to do with financial soundness than with the need to carry water for an activist agenda.[7]

The Congressional Black Caucus and the Congressional Hispanic Caucus constantly pressed for more. And Fannie Mae and the Clinton administration were eager to comply. At a 2005 swearing-in ceremony for new members of the Congressional Black Caucus, including Senator Barack Obama, Fannie Mae CEO Daniel Mudd (named interim CEO in 2004 and confirmed in 2005) declared: "I humbly ask you to help us and help me. . . . If there are areas where we could do better, we'd like to hear it from our friends, and I'd be so bold as to say our family, first." He went on to explain, "In many ways . . . you are also the conscience of Fannie Mae." This sort of spineless pandering to minority special interests is all too common among guilty white liberals of the Clinton generation. And so the looting and the payoffs continued.

Fannie and Freddie also lavished contributions on activist groups, which in turn served as a shield against congressional scrutiny. (The merest hint of opposition to the activist agenda would trigger a hurricane of public demonstrations and loud charges of "racism.") They gave generously to ACORN and the Center for Community Change, both of which were aggressively pushing for expansion of the Community Reinvestment Act. They also gave to Jesse Jackson's Rainbow-PUSH Coalition and the Congressional Black and Hispanic Caucus Foundations. Large sums of money flowed unhindered into the coffers of the Left.[8]

The political seduction of Fannie Mae was quickly accomplished. But its march to financial oblivion began with the rise of James A. Johnson as CEO. The tall Minnesotan was in many ways the consummate D.C. insider, with a long pedigree in the liberal wing of the Democratic Party. His father had been a Minnesota state legislator, and the younger Johnson cut his teeth in the Eugene McCarthy and George McGovern campaigns. He went on to work for his fellow Minnesotan Walter Mondale in the Senate, followed him to the White House as the vice president's executive assistant during the Carter years, then chaired Mondale's unsuccessful 1984 presidential campaign. (Later he would become a key fund-raiser and close adviser in turn to John Kerry, Hillary Clinton, and Barack Obama.) He then went into business with his friend Ambassador Richard Holbrooke—another Clintonian figure—to form Public Strategies, a Washington, D.C., consulting firm. One of their early clients was Fannie Mae.

Johnson became chairman and CEO of Fannie Mae in 1991. He had never run a company and had done little outside the world of politics. As CEO, he went around the country meeting with banks and lenders, which were in effect Fannie's clients. As one friend described it, "Jim wanted to go out and learn the industry."[9]

Johnson really didn't need much convincing by the activist groups. As a good liberal, he already thought it was his mission to "transform the housing finance system" and promised to ensure that "every American who wants to get a mortgage will have their loan approved." He announced that Fannie Mae would commit itself to purchasing $1 trillion in loans for lower-income and minority groups and new immigrants. As the *New York Times* explained, "The initiative partly reflects Mr. Johnson's idealistic view of Fannie Mae as an engine for social improvement and his long history of civil rights activism." Here is a question not asked by the *Times*: since when did a long history of civil rights activism become a qualification for running a $350 billion government-backed financial agency?[10]

Of course, in order to fulfill this pledge, Johnson had to move Fannie Mae away from its traditionally conservative underwriting principles. He pressed banks to ignore or inflate credit scores, which he condemned as "mechanical systems and arbitrary numbers." As Johnson said, "It seems we can make A's out of B's and C's in many circumstances" by ignoring relevant factors such as job losses.[11]

Meanwhile, the Clinton administration was taking steps to place the GSEs more firmly under government control. In 1992, Congress passed a law requiring the U.S. Department of Housing and Urban Development (HUD) to make sure that national housing goals (as determined by the federal government) were being met every four years. In effect, this allowed HUD to set goals for Fannie Mae. The person responsible at HUD for setting these goals was Roberta Achtenberg, who could be counted on to press relentlessly to align Fannie and Freddie with activist goals. In keeping with the Clinton administration's ambitious housing agenda, HUD Secretary Henry Cisneros declared in December 1995 that 42 percent of the mortgages

Fannie Mae bought should now be for low- and moderate-income families.

This was a radical departure from how Fannie Mae had been run in the past. And it would dramatically affect the mortgage market. Since Fannie Mae bought mortgages from lenders, this meant in turn that the agency would be encouraging more loans to these groups. Accordingly, it had to loosen its standards. As a result, by 1997, it for the first time was purchasing mortgages with only 3 percent down. Barry Zigas, who headed these efforts at Fannie Mae, acknowledged, "There is obviously a limit beyond which [we] can't push [the banks] to produce." But clearly he didn't think the agency was there yet.[12] The same year, Cisneros agreed to let Fannie and Freddie meet this goal in part by buying subprime mortgages from lenders for the first time. Fannie Mae was now in the subprime mortgage business.[13]

Johnson was willing to do whatever he could to fuel the purchase of subprime mortgages. When lenders sold mortgages to Fannie Mae, they made certain guarantees about their quality. For example, a lender might say that the average credit score of the mortgages that were bundled together and offered to Fannie Mae was 660. If Fannie Mae purchased them and discovered after the sale that this was not true, part of the agreement was that the lender would have to buy them back from Fannie. But when it came to some of the subprime mortgages for low-income buyers, the credit scores were terrible and often misreported. Johnson loosened those standards and said that he wanted to reduce the number of buybacks, meaning that even if lenders had not provided accurate information on how safe the mortgages were, he would not insist that they buy them back. According to Johnson, "The most important effect of the changes will be an increase in affordable housing lending nationwide by a reduction in lenders' worries that they will be forced to buy back riskier loans." The idea that it might be a good idea for subprime lenders to know that they would be forced to buy back riskier loans to keep them honest didn't seem to cross his mind.

"The first half of this decade has been a period of exciting new activity on the affordable lending front," Johnson said in 1996. "Our customers [banks and mortgage companies] stepped up to the challenge of providing more access to mortgage credit for low- and moderate-income families, minorities, new immigrants and other Americans with special housing needs."[14]

Fannie Mae embraced its new activist and social engineering mission. Following the lead of ACORN and other activist groups, Fannie Mae encouraged lenders to be flexible, advising them that when it came to assessing assets and income, "primary income may be supplemented by income from family members who are on disability or work off the books." Fannie Mae also encouraged lenders to give mortgages to "certain nonpermanent immigrants" (what some might call illegal immigrants), noting that in these instances, "many times the borrower would have more than one social security number."[15]

Fannie also helpfully noted that some of these undocumented immigrants might have unusual credit histories with gaps in income due to the fact that they "go home, then come back and sign on with another employer" and might also take "the annual customary 3-month return to Mexico at Christmas time."[16]

Fannie partnered with activist groups such as National Council of La Raza, a Latino civil rights organization, and Chicanos por la Causa, a Latino activist group, to help them find immigrant mortgage applicants.

Despite these efforts, the full-court press by housing activists continued. Granted numerous concessions, they demanded even more. Unsatisfied with Cisneros's mandate that 42 percent of mortgage purchases go to the poor and middle class, his successor, HUD Secretary Andrew Cuomo, commissioned the Urban Institute to study Fannie and Freddie's underwriting guidelines. Formed in 1968 with the blessings of President Lyndon Johnson as a means of monitoring the progress of Great Society programs, the Urban Institute was a reliably liberal group that had argued, for example, that the 1992 riots in Los Angeles (sparked by the police beating of the black motorist Rodney King) had been the result of social and economic inequality.

Now the Urban Institute concluded that Fannie and Freddie—despite all they had already done to lower credit barriers—had racially biased standards: "the GSEs' guidelines, designed to identify creditworthy applicants, are more likely to disqualify borrowers with low incomes, limited wealth and poor credit histories; applicants with these characteristics are disproportionately minorities." Imagine that! "Informants said that some local and regional lenders serve a greater number of creditworthy low-to-moderate income and minority borrowers than the GSEs, using loan products with more flexible underwriting guidelines than those allowed by Fannie and Freddie." [17]

The Urban Institute, it should be noted, is itself heavily underwritten by the Fannie Mae Foundation, Bank of America Foundation, and American Express Foundation, among others.

Martin Luther King III—the son of the fallen civil rights leader—also joined the chorus of those who argued that Fannie and Freddie were "falling short" in their responsibilities. [18]

In 1999, Andrew Cuomo announced that he was increasing the agencies' affordable housing goals, justified in part by the untested theory that a government mandate was a healthy way to "stimulate" the housing economy. "This action will transform the lives of millions of families across our country by giving them new opportunities to buy homes," he announced. "It will strengthen our economy and create jobs by stimulating more home construction . . . and it will help reduce the huge home ownership gap dividing whites from minorities and suburbs from cities." [19]

Cuomo dramatically raised the limits on the size of FHA loans and cut the down payment requirement to 3 percent. Cuomo also strong-armed companies to offer mortgages not just for low-income but for very-low-income applicants. As a result, Fannie's investment in subprime mortgates rocketed from $1.2 billion to $81 billion by 2003. Then he also raised the portion of Fannie and Freddie loans for poor and middle-class mortgages to 50 percent. What had once been large government-backed entities purchasing safe and reliable mortgages were now using half of their funds to purchase increasingly risky subprime mortgages. And Cuomo dramatically hiked the mandates that they buy mortgages in what were regarded as very-low-income groups.

Cuomo's rationale was plainly based on race. In laying out his new goals, he announced in his report, "GSE presence in the subprime market could be of significant benefit to lower-income families, minorities, and families living in underserved areas." Cuomo's aide William Apgar was even more explicit in the *Washington Post*: "We believe that there are a lot of loans to black Americans that could be safely purchased by Fannie Mae and Freddie Mac if these companies were more flexible." By the end of Cuomo's tenure in 2001, Fannie was buying mortgages with no money down at all.[20]

James Johnson and the Fannie Mae leadership did not object to those goals. Indeed, they eagerly embraced them. "We need to push into these underserved markets as much as we can," explained David Glenn, president and chief operating officer of Freddie Mac. Timothy Howard, Fannie's chief financial officer, explained that "making loans to people with less-than-perfect credit" is "something we should do." Fannie developed a line of "flexible" mortgages, including 100 percent financing and requiring as little as $500 to cover costs.

Jesse Jackson loved the new relaxed standards at Fannie and Freddie. He believed that buying up large quantities of subprime mortgage-backed securities was a good idea, too. In a book on personal finance that he cowrote with his son, Representative Jesse Jackson, Jr., titled *It's About the Money!*, they explained how Fannie Mae could "increase the liquidity of the nation's mortgage market by buying mortgages from lenders and repackaging them into securities that are then used as collateral for securities sold on Wall Street. This is the origin of the 'mortgage-backed' securities industry. Because of this arrangement, the lender gets payment immediately and can go ahead and make more mortgages, thus fueling homeownership in this country."

Writing in 1999, the Jacksons explained that

> in the last five to ten years, Freddie Mac and Fannie Mae have begun offering increasingly flexible loan products to meet the needs of low-income people and minorities. The overall homeownership rate in America is at an all-time high of 66.7 percent. . . . Thanks to these initiatives minorities' homeownership rates are growing twice as fast as those for whites.
>
> Many people cite difficulty in accumulating the down payment as their major obstacle in homeownership. However, thanks to programs run by Freddie Mac and Fannie Mae, it's now possible to purchase a home with only a 3 percent down payment and total up-front costs of around $4,000.[21]

Fannie and Freddie, as well as the Clinton administration, knew at some level that this was a recipe for disaster. There was plenty of evidence that by pushing into "underserved markets" these taxpayer-backed enterprises were being encouraged to purchase large quantities of mortgages from people who had lousy credit histories. In fact, at the time Freddie Mac conducted a study on the credit histories of 12,000 Americans and discovered that a higher percent of African Americans with incomes of $65,000 to $75,000 had 'bad credit' than whites with incomes below $25,000. The study also found that half of black borrowers and a third of Hispanics had "bad" credit records, meaning delinquent loans or bankruptcy.[22]

What did this mean? It meant that Fannie and Freddie were buying an increasing number of lower-quality loans to meet the Clinton administration-imposed goals and standards. In 2000, Freddie bought $18.6 billion in subprime loans. (Fannie did not disclosure its figures.) By 2004, Freddie and Fannie's purchases of subprime mortgages had risen tenfold, just to meet the affordability goals. The two agencies together purchased 44 percent of mortgages originated in the subprime market.

There were, of course, plenty of warning signs that this was all headed for disaster. A 2001 HUD report warned of the consequences: "Given the very high concentration of these loans in low-income and African American neighborhoods, the growth in subprime lending and resulting very high levels of foreclosure is a real cause for concern."[23] In 2004, HUD again expressed concern that the push to lower lending standards had created an irreconcilable tension between meeting the agencies' goals and responsible business practices. Even though alarm bells were going off in some quarters, many individuals in the Bush administration shared the home-ownership goals of Clinton and the housing activists. President Bush, encouraged by HUD Secretary Alphonso Jackson, committed to dramatically increasing minority housing goals and making mortgages more affordable by offering more 100 percent loans.

From 2005 to 2007, based on Cuomo's new standards for Fannie and Freddie, these two taxpayer-backed entities bought $1 trillion in subprime and Alt-A loans, an astonishing 40 percent of their total mortgage purchases. (Alt-A loans are for individuals with somewhat below average credit scores.) Fully 10 percent of Freddie loans purchased by 2004 were by home buyers who had abysmal credit scores of less than 620 (660 is considered subprime). Those numbers jumped to 14 percent in 2005 and an astonishing 30 percent in 2007. In all, Fannie bought 57.5 percent of its loans from this category during the same period.[24]

Fannie Mae also plunged fearlessly into the world of CRA loans. Johnson set a goal of purchasing or securitizing more than $530 billion in CRA loans to poor borrowers with bad credit. And he began offering mortgage products to make that possible, including so-called Community 100 mortgages, which included a 100 percent mortgage, and Fannie 97 mortgages, which required just 3 percent down. Johnson also had the company's investor trading desk create "customized mortgage-backed securities" for CRA-qualified loans.[25]

Johnson put pressure on banks to make high-risk loans so that he could buy the mortgages and meet his quota. If they were not writing such loans, he assumed it was because of racial bias. As he explained it, "If banks don't make home loans in underserved communities because of bias of their loan officers or underwriters, we can't buy loans. But if our clients generate more loans in their communities, we'll have more loans to buy, which helps us meet our business goals."[26]

The term "underserved communities" is a term of art for minority neighborhoods that represent high concentrations of risk from a lender's perspective. By using this term, however, the government official gave the impression that it was the fault of the banking industry that those communities were not being properly "served"—in other words, it was a failure of the marketplace, not a rational judgment by the market itself. The government was therefore justified in stepping in and making sure the "market" functioned properly.

One can perhaps see how this attitude might produce a major blind spot, leading in turn to a massive distortion in the housing economy. This also helps explain why the officials who ran this insane racket were able to accept enormous salaries, bonuses, and retirement packages with a clear conscience. They had done the Lord's work, after all. Why shouldn't they be handsomely rewarded?

In 1999, James Johnson left Fannie Mae. In recognition of his service, Johnson was awarded a postemployment contract worth more than $390,000 per year, along with two employees as support staff, plus a car and driver. This was on top of a $71,000-per-month pension—that's right, *per month*!—that he would get for life (he was then fifty-six years old—do the math yourself). Shortly thereafter, Johnson was named to the board of Goldman Sachs and joined the compensation committee. There he would help set the salary for Henry Paulson, who would later become Treasury secretary in the Bush administration.[27]

Johnson's replacement at Fannie Mae proved to be even more reckless in trying to achieve his social engineering goals through a company backed by U.S. taxpayers.

Franklin Raines had an impressive life story and résumé. The son of an African-American janitor from Seattle, he graduated from Harvard and served in the Carter administration at the Office of Management and Budget. Unlike Johnson, he had real experience in the world of finance. After leaving the Carter administration he went to work at the Wall Street investment house Lazard Frères, where he rose to general partner and became nationally known as the first black man to head a Fortune 500 company. He then joined the Fannie Mae board and soon became vice chairman. Subsequently he served as President Clinton's budget director.

A true child of the sixties, Raines promoted the idea of corporate citizenship as a means of bridging the gap between liberal goals and capitalist realities. Recalling his senior year in college in 1971 in a speech at the University of Connecticut, he remarked on how much attitudes had changed since then about the proper role of business in society. "Most people would have said it was the job of individuals or government—not corporations—to improve society. Today, however, we live in a society where the boundaries of business and government are beginning to blur. We have come to accept that government cannot and should not do it all. Corporations have a greater social responsibility today." He spoke often and with passion about the fact that home ownership was "unevenly distributed in society." (Like other fair housing advocates, Raines apparently believed that homes are "distributed," not earned and purchased by buyers.) Speaking on college campuses, Raines would often quote W. E. B. Du Bois's view that the size of a person's home is an index of his or her condition and lamented the fact that in America, "minorities have yet to achieve parity in home ownership."

Shortly after Raines took over as CEO, Fannie Mae met its goal of providing $1 trillion for underserved communities in the United States. Indeed, it met it ahead of schedule. Raines upped the goal by another $2 trillion, promising to help 18 million more families become first-time home owners.

Raines was quite clear and direct about the fact that this was a massive effort in social engineering backed by the federal government. As he told students at the University of Connecticut in 2001, Fannie Mae was "a mediating structure that bends the financial system to create homeowners." It was a very revealing statement. Far from his seeing Fannie Mae as a huge distorting factor in the mortgage market, its operation was consistent with his philosophy of do-good capitalism and its notions about corporate citizenship. In the view of Raines and other liberals of his generation, companies (including and especially banks and other lenders) have an important role to play in promoting a good society. Since the operations of the market alone do not produce the desired social results, the market must be jimmied and coerced using the power of the federal government. Thus he proudly explained that Fannie Mae supported more home ownership than "the free market alone would."[28]

As Fannie and Freddie gobbled up subprime loans, they created greater demand from mortgage originators and brokers, who knew they could sell them. It was in market terms a rational response. Wall Street found such instruments very profitable. The fact that Fannie and Freddie were heavily into the market indicated that, at some level, the system was safe. The federal guarantee kicked in to tip the scales.

The Clinton administration's decision to ease credit requirements and start gobbling up subprime loans did raise concerns, however—even at the *New York Times*. "Fannie Mae Corporation is easing the credit requirements on loans that it will purchase from banks and other lenders," the paper reported in 1999. "This will encourage those banks to extend home mortgages to individuals whose credit is generally not good enough to qualify for conventional loans." What that meant, according to the *Times*, was that Fannie Mae "is taking on significantly more risk, which may not pose any difficulties during flush economic times. But the government-subsidized corporation may run into trouble in an economic downturn, prompting a government rescue similar to that of the savings and loan industry in the 1990s."[29]

It used to be that Fannie and Freddie played the role of gatekeeper. With their strict underwriting standards—minimum down payments, mortgage insurance requirements, thick application forms, and detailed credit histories—they would only buy loans meeting strict requirements. But with the new mandates for affordable housing and loosening of underwriting standards, most of these requirements were melting away. Soon Fannie Mae CEO Jim Johnson was "linked at the hip" with subprime lenders such as Countrywide, which had issued a massive number of loosely underwritten loans.

Indeed, in any given year, Countrywide might account for up to 30 percent of all the loans that Fannie Mae bought. Johnson would invite Countrywide CEO Angelo Mozilo—later named by CNN as one of the "Ten Most Wanted" culprits of the 2008 financial collapse—to speak at Fannie Mae retreats for senior executives and sales teams. Johnson, in turn, would frequently fly on the Countrywide corporate jet. When Johnson left Fannie Mae and became chairman of the Kennedy Center in Washington, D.C., Mozilo would frequently sit next to him in the VIP box to watch Broadway shows. It became a joke in the mortgage industry that Countrywide "was really just a subsidiary of Fannie Mae."[30] As Mozilo himself put it, "If Fannie and Freddie catch a cold, I catch the f——g flu."[31]

Mozilo, the son of a Bronx butcher who attended Fordham University, cofounded Countrywide Credit Industries in 1978. By 2000, he had built the company into the largest lender in the country—with vital help from the federal government.

Often called "the Tan Man" because of his bronzed complexion, Mozilo saw his efforts to sign up subprime loans as a sort of social crusade couched in language that might have come from the pages of *The Nation* or the lips of the filmmaker Michael Moore. "We [at Countrywide] have a stated mission to make a difference in the lives of American families, to get low-income people into housing," he said. "I have nine grandkids. There is too much division when you have haves and have-nots. You don't have peace in the world." By giving out subprime mortgages, no-money-down mortgages, and even mortgages that allowed the buyer to skip a payment, the Tan Man would save the world.[32]

Countrywide was, for much of the housing boom, the largest lender to Hispanics and blacks in the country, a fact Mozilo often mentioned in his speeches. He offered mortgage-writing services in six languages for potential home owners who couldn't speak English, offered 103 percent mortgages that covered closing costs, and even proposed a "multicultural approach" to matters of credit history. As he explained, his company's core customers were those in "multicultural markets."[33] Fannie Mae loved this because it conformed to its own federally mandated goals.

Mozilo also opened a trading desk to buy and sell jumbo loans and CRA loans. Banks that were being shaken down and needed CRA credits could buy them from Mozilo. "On our first deal we sold $12 million in Alaska loans to Goldome Bank, which needed loans for CRA purposes," bragged Jonas Roth, who ran the operation. "We made double the profit we could've made by selling to Goldome. We made a killing. Goldome needed those CRA credits. We knew we had something there."[34]

Mozilo was seen as a hero in liberal quarters for his efforts to help the downtrodden. In 2000, the Hispanic magazine *La Opinión* named Countrywide "Corporation of the Year" for its commitment to diversity and aggressive lending to Latinos. Mozilo was feted at an awards presentation at Union State in Los Angeles. The liberal, Berkeley-based Greenlining Institute likewise praised him for his commitment to minority home ownership. "Mozilo said the right things about minority home ownership rates, and that's a laudable thing to do," said Bob Gnaizda.[35] The National Housing Conference piled on with a lifetime achievement award "in recognition of his longstanding commitment to reducing the barriers of homeownership for lower-income and minorities."

In February 2003, Mozilo was invited to deliver the prestigious John T. Dunlop Lecture at Harvard University. In a speech that could have been written from the liberal corner of the Harvard faculty lounge, Mozilo spoke of the need to "push for greater diversity in homeownership" and deplored the fact that the poor and minority communities did not have the same ownership rate as whites. The gap, he said, "remains intolerably too wide."

Mozilo praised the CRA and argued for a multicultural approach to credit. "We must look for ways to capture 'alternative' payment histories and to properly factor in cultural differences in credit, income and spending habits, so that we can say 'yes' to borrowers who have the ability and willingness to make their mortgage payments. Credit scores must not be a dominant factor for assessing risk. Nontraditional factors such as rent and utility payment history should be imbedded in the automated underwriting process. . . . Let's look for every possible reason to approve applicants, not to reject them." He also praised "flexible underwriting techniques [that] are continuing to fuel a record period of growth" in home purchases and lauded the role that the CRA was playing. To deal with the lingering "gap" in minority home ownership, he proposed "elimination of down payment requirements for low-income and minority borrowers." The traditional practice of requiring 10 percent down for a mortgage "adds absolutely no value to the quality of the loan," Mozilo claimed.

The Countrywide approach was multiculturalism run amok. As Rodolfo Saenz, the company's executive vice president for emerging markets, put it, "All of the nations' top ten mortgage originators and servicers have responded to the need for affordable, flexible financial programs." Countrywide and others "offer zero- and low-down payment loans and underwriting guidelines that recognize diversity and cultural differences in the way minorities and immigrants may view and conduct their personal financial situations." Defaults and poor credit histories were now "cultural differences"? Not only was this a dangerous game to be playing with the country's financial future, the whole notion was highly insulting. It was in effect saying that certain cultural groups simply couldn't act responsibly with their money, the rules needed to be bent.[36] But all of this was completely consistent with what was going on at Fannie Mae and Freddie Mac.

Then Mozilo made a bold commitment, not only to continue aggressive lending to underserved communities through the company's "House America" program but to sign an astonishing $600 billion in new loans on top of the $100 billion Countrywide had already written over the past twenty-two months. He went on to thank the "partners" who were helping him in this noble enterprise: Fannie Mae, Freddie Mac, the Congressional Black Caucus, the National Council of La Raza, and the AFL-CIO.

How did Mozilo, an obviously shrewd businessman, build a business on risky subprime loans? He knew they were risky and dangerous. In one e-mail to his colleagues which the Securities and Exchange

Commission recently released as part of a civil suit it is filing against him, Mozilo wrote, "In all my years in business I have never seen a more toxic product." In another he wrote that the no-money-down mortgage was "the most dangerous product in existence and there can be nothing more toxic." [37] Mozilo originated these mortgages, but he didn't keep them. Instead, he turned around and sold them to others. And his biggest customer was the taxpayer-backed, government-supported enterprise Fannie Mae. Moreover, he was doing exactly what HUD and Fannie Mae wanted him to do. They were, in fact, his enablers.

In his speech at Harvard, Mozilo acknowledged that he couldn't do any of this without the financial support of Fannie Mae. "I specifically and especially recognize Franklin Raines and his entire team at Fannie Mae for providing a great deal of the resources that have made it possible for us to achieve our House America objectives."

A year later the *New York Times* reported how Fannie and Freddie had propelled Mozilo to the top. "By buying up his mortgages and thus freeing up his capital to solicit even more business, Fannie and Freddie are a big reason Mr. Mozilo has driven Countrywide past the Contigroups and the Wells Fargos to the top of the mortgage heap." [38]

Mozilo had to worry substantially less about risk because government-backed entities were reducing the risks for him. He also had little incentive to ensure that the mortgages he was writing didn't go bad. He therefore mastered the art of pushing NINJA (No Income, No Job, No Assets) loans, "liar loans" (which relied on an applicant's stated income rather than proof of income), and even mortgages that allowed the home owner to skip payments now and again. It was quite an arrangement: sign up billions in mortgages, and then sell most of them off to a taxpayer-guaranteed enterprise.

When the Bush administration tried to cap the size of loans that Fannie Mae could buy from the likes of Countrywide at $417,000

each, Mozilo became the populist fighting against Republican plu-
tocrats. "They're ideologues," he said of the Bush administration.
"They never sat around the kitchen table with their parents, trying to
figure out how to make that month's rent."[39]

Given its nature, Countrywide was as much a political operation
as a business. So in addition to saying the right things, Mozilo tossed
cash to the right people—those who would support his interests and
promote the goals of Fannie and Freddie, which were so important
to his business model. Without Fannie and Freddie, Countrywide
would be a fraction of its size. So Mozilo gave ample campaign contributions to both political parties but proved to be especially generous to those candidates like former HUD Secretary Andrew Cuomo
who shared his views. He also gave generously to the Congressional
Black Caucus (the "conscience" of Fannie Mae).

Countrywide also gave $2 million to a program called Hogar
(Spanish for "hearth") set up by the Congressional Hispanic Cau-
cus Foundation to increase Hispanic ownership. Hogar pushed for
relaxed lending standards and echoed Mozilo's view that since many
Hispanic families had more than one job and little savings, they
needed flexible loans. Other financial supporters of Hogar included
Fannie and Freddie, Washington Mutual, Ameriquest, and New
Century Financial.

Hogar collaborated with Freddie Mac to study Latino home ownership and found that rates were on the rise due to the "new flexible
mortgage loan products" that Countrywide and others were offering.
The report's recommendation, needless to say, was for the further
easing of standards. Countrywide and Fannie Mae were no doubt
very happy.

Hogar worked with subprime lenders such as Countrywide to
find Latino applicants who might benefit from these "flexible mortgage loan products," which in turn could be sold to Fannie Mae. The
resulting loans included a $330,000 no-money-down mortgage for a

part-time electrician from Bolivia. "I said this is too good to be true," he recalled in the *Wall Street Journal*. "I'm 23 years old, with a family buying my own house." But when work slowed down he lost his home. Another $370,000 mortgage went to a $28,000-a-year office manager.[40]

Mozilo put politically well-connected people on his board, including former Clinton HUD Secretary Henry Cisneros and Kathleen Brown of Goldman Sachs, who just happened to be the sister of former California Governor Jerry Brown. Speaker of the House Nancy Pelosi's son was on the payroll, while at the same time holding another job for a company called InfoUSA. Mozilo also created a secretive program to offer below-market mortgages to political allies who supported the interests of his company and the mission of Fannie and Freddie. Essentially, these "Friends of Angelo" could finance high-cost mortgages at low rates without closing costs or other expenses.

Senator Christopher Dodd, the Chairman of the Senate Banking Committee, which regulates mortgage lending, received two special mortgages from Countrywide in which points were shaved and Dodd saved thousands of dollars. The word must have hit the street, because Senator Kent Conrad called Mozilo directly and asked him for a mortgage. Conrad received a special rate that saved him $10,000 a year on a $1 million vacation home in Delaware. He also received a mortgage for an eight-unit apartment building, even though Countrywide had a policy of not providing loans for such buildings. According to an internal Countrywide e-mail written by the Tan Man, the exception was "due to the fact that the borrower is a senator." Others who received special treatment included former Clinton Health and Human Services Secretary Donna Shalala, ex-Fannie Mae CEOs Jim Johnson and Franklin Raines, and Johnson's former business partner Richard Holbrooke, who is now Barack Obama's special envoy to Afghanistan.[41]

Another large subprime lender, Ameriquest, adopted a similar political strategy designed to shelter it from criticism and advance the agenda of Fannie Mae. CEO Roland Arnall, who was a cofounder of the Simon Wiesenthal Center and the Museum of Tolerance, was a political contributor to both political parties. He gave hundreds of thousands of dollars to California Governor Gray Davis and then to his successor, Arnold Schwarzenegger. After favoring Democrats for decades, he then became a large donor to President George W. Bush. He also gave generously to Hogar and to members of the Latino Caucus in the California State Assembly.

Ameriquest gave hundreds of thousands of dollars in donations to the Leadership Conference on Civil Rights, the largest and oldest civil rights coalition in the United States. (The Leadership Conference and many member organizations were staunch supporters of Fannie Mae.) When Arnall was appointed ambassador to the Netherlands by President George W. Bush in 2006, the LCCR came vigorously to his defense after criticisms were leveled against Ameriquest for its lending practices.

Like Countrywide, Ameriquest gave generously to Hogar. Arnall also put politically connected individuals on its board, including the Clinton Justice Department official and current governor of Massachusetts Deval Patrick. The company forged ties with now–Los Angeles Mayor Antonio Villaraigosa. The onetime member of the group Chicano Student Movement of Aztlán (MEChA), which calls for the Latino "reconquista" of the southwestern United States, Ameriquest became a longtime patron of Villaraigosa. It donated $75,000 to his political action committee when Villaraigosa was speaker of the California State Assembly from 1998 to 2000. When he left the legislature, Ameriquest hired him as a consultant. (Villaraigosa once said he was not a lobbyist for the firm; he just provided "strategic thinking and problem solving.") Later, as mayor, he flew on the Ameriquest corporate jet to civil rights events, such as the funeral of the civil rights pioneer Rosa Parks. (The mayor reimbursed Ameriquest less than $1,000 for the flight, or less than 10 percent of the actual cost.)[42]

This political strategy was replicated by other subprime lenders. New Century Financial Corporation donated generously to the Hispanic Congressional Caucus. A subprime lender named Delta Financial Corporation placed Maggie Williams, Hillary Clinton's campaign manager and longtime family friend, on its board. She earned $200,000 before the firm went bankrupt. Williams subsequently defended the discredited lender, arguing, "I joined the board because I . . . understood that the subprime option, for all its challenges, was the only chance for many people to own a home." Delta charged an average interest rate of 10 to 12 percent on home mortgages, at a time when the typical rate was 6 percent.

Another giant lender that adopted this strategy was Washington Mutual (WaMu). Unlike the other lenders, WaMu was a bank—and it became the sixth largest in America by pushing the same products to the same customers as Countrywide and Ameriquest were. Its lending standards were no higher than theirs. A mariachi singer in California claimed a six-figure income to get a mortgage and provided a picture of himself in a hat with his guitar as proof; that was good enough for WaMu.

WaMu head Kerry Killinger embraced the activist agenda. Thus a vice president of Washington Mutual served as chair of Hogar's advisory committee. And Killinger was committed to moving into new markets and exhibited a pronounced social liberalism in his attitudes about finance. His goal was to transform the company into the "Wal-Mart of banking" by catering to lower- and middle-class consumers that other banks deemed too risky. The company even established internal quotas in its lending to poor and minority applicants. There was "a company policy mandating that its performances within this demographic in a given market at least matches the bank's overall position in the market."

The company offered 100 percent financing with the customer coughing up as little as $500 to move into a new home. It even had plans to develop a special task force to offer mortgage financing to Native Americans.

"Affordable housing and lending is front and center in terms of our strategy," said Killinger. In 2003, he struck a deal to sell Fannie Mae $85 billion in "affordable mortgages."

WaMu played the game with activists and committed hundreds of billions of dollars in loans for "low- and moderate-income borrowers, and minority borrowers." It pledged $375 billion to that market and committed nearly $1 trillion in mortgages to those with credit histories that "fall outside typical credit, income or debt constraints." The company received the 2003 CRA Community Impact Award for its Community Access program for low-income people.[43]

Whether Mozilo and the others in the subprime industry actually believed their own rhetoric is hard to say. What we do know is that by using the currency of civil rights, income inequality, and multiculturalism, they developed sufficient political cover to allow them to continue their egregious lending practices. They could avoid criticism and keep regulators at bay. But it also showed that community leaders and so-called civil rights activists could be bought off relatively easily. And Fannie Mae made this explosive growth possible, because the taxpayer-backed enterprise *wanted* to buy up risky mortgages. And the subprime lenders were all too willing to sell.

The explosion of lending to poor and minority applicants provided healthy profits for these companies. And the house of cards stood, at least for a while. But when the subprime mess exploded, Mozilo and the others found themselves abandoned by those who had praised them only a few years earlier. *ABC World News Tonight* featured a story that began with a screen-filling photo of Mozilo: "This may well become the deeply tanned face of the mortgage mess. The face belongs to Angelo Mozilo, the once-celebrated CEO of Countrywide, now facing allegations of predatory lending and rapacious greed."

Countrywide was bought by Bank of America after the company suffered serious losses in its mortgage division. Ameriquest was bought by Citigroup after suffering severe losses. New Century filed for Chapter 11 bankruptcy. Just a couple of years after Washington Mutual pledged $325 billion in lending to low-income borrowers, it was taken over by the Office of Thrift Supervision and placed into receivership under the Federal Deposit Insurance Corporation. Mozilo would later claim that he was an innocent victim in all of this. As the *New York Times* reported: "At a conference sponsored by the Milken Institute about two weeks ago [Mozilo] explained that borrowers forced lenders like Countrywide to lower their mortgage standards. The industry faced special pressure from minority advocates to help people buy homes, he said. Now, the government must help by increasing loan limits at government-sponsored enterprises like Fannie Mae and Freddie Mac, he added."[44]

Mozilo was partly correct: housing activists, abetted by their counterparts in the Clinton administration, did pressure banks and other lenders to lower standards. But Mozilo and the other subprime lenders were hardly victims; they were more like willing accomplices in a massive social engineering scheme that has devastated large segments of the American population.

Fannie and Freddie were hijacked by activists. That is what activists do: they try to push their agenda. Subprime lenders such as Countrywide gorged on profits by selling toxic loans to these taxpayer-supported entities. And that is what companies do: they seek out profits. But with taxpayers on the hook for trillions of dollars in potentially toxic loans, the bigger question to be asked is, what were the politicians in Washington doing? The unfortunate answer: all too much.

THE GOLDEN TROUGH

**How Liberal Politicians Used Fannie and Freddie to Rig
the Real Estate Market While Lining Their Own Pockets**

Today Fannie and Freddie are behemoths of debt and, as such, prime incubators of the economic crisis. If you add together the mortgages they hold and the mortgages they have sold to investors around the world and on which they have offered a payment guarantee, these two companies hold potential liabilities of some $5 trillion. (This is a global problem: an estimated $1.45 trillion of these debts are held overseas.) In effect, these two government-sponsored entities have liabilities equaling about half the current U.S. national debt. To make matters worse, much of that debt is in the form of mortgage-backed securities (MBSs), and a significant proportion of that is in subprime loans.

This was not a sudden train wreck that took people in Washington by surprise. It has long been known that Fannie and Freddie were teetering on the tracks and possibly ready to go off the rails. Why was nothing done? The short answer is that Fannie and Freddie served the ideological interests of the activists. And politicians in Washington turned the other way because they and their allies could profit handsomely from the scheme.

Fannie and Freddie have been exempt from the normal oversight and regulatory requirements of privately owned financial companies. As William Poole of the Federal Reserve Bank of St. Louis put it, "Capital on the books of Fannie Mae and Freddie Mac is well below the levels required of regulated depository institutions." Commercial banks, for example, are required to hold equity capital and subordinated debt of a bit under 11 percent of total assets. But for Fannie and Freddie, "The core capital requirement is 2.5 percent of on-balance sheet assets and 0.45 percent of outstanding mortgage-backed securities and other off-balance sheet obligations." Needless to say, U.S. taxpayers are on the hook for all of it.[1]

At the same time, members of Congress from both political parties, federal bureaucrats, and financiers on Wall Street served as willing enablers, allowing this to happen and profiting handsomely along the way. The companies' management positions became a dumping ground for failed candidates and ex-bureaucrats, who typically earned huge salaries. Fannie Mae, for example, pays its twenty-one top executives more than $1 million apiece. Three of these are in-house lobbyists whose main job is to prevent Congress from pushing reforms, threatening its special status, or questioning its financial soundness.[2]

The activities of Fannie and Freddie, which are little covered by the press, have been shrouded in mystery for years. But the agencies, with their enormous pots of cash, were in effect captured by politicians on Capitol Hill, who turned them into a kind of private piggy bank, using them to finance their campaigns, underwrite their pet causes, and ensure soft landings for a revolving cast of Washington insiders. Beltway elites made large sums of money consulting for, lobbying on behalf of, or simply serving on the boards of these large entities, while doing very little in return.

The boards of both are sprinkled with defeated candidates, ex–White House aides, and former government bureaucrats. President Clinton, for example, placed Rahm Emanuel, Harold Ickes, Jack Quinn, and Eli Segal on the boards. Emanuel was a political adviser, Quinn was a lawyer and aide to Al Gore, Ickes was a political activist and lobbyist connected to Hillary Clinton, and Segal was Clinton's campaign chief of staff. None of them had the remotest qualifications to help run and direct these large financial entities. When Webster Hubbell, the former associate attorney general who was at the heart of the Whitewater affair, was facing financial troubles, White House officials called Fannie Mae and secured a job for his son.[3]

The same game was played by Republicans. George W. Bush found a slot for the wife of former Michigan Governor John Engler and a former regional reelection campaign manager on the board, but he eventually ended the practice of using the boards for patronage—the first president to do so.

Serving on these boards could be outrageously profitable. Consider the case of Rahm Emanuel, currently chief of staff in the Obama White House. After his stint as Clinton's political director, the president appointed him to the Freddie Mac board in 2000. He served for fourteen months, attended just seven board meetings, and sat on no committees. (According to fellow board member Neil Hartigan, the former Illinois attorney general, Emanuel's "primary contribution was explaining to others on the board how to play the levers of power.") In exchange he received more than $320,000 in compensation. In short, he made approximately $46,000 an hour for his service on this government-sponsored company.[4]

Democrats in Congress, and many Republicans too, nodded with approval like bobble-headed dolls at anything Fannie or Freddie wanted from the federal government. The two GSEs passed huge sums of money around on Capitol Hill, effectively buying large segments of the Washington establishment who would champion their mission and making people from both parties very wealthy. According to Gerald O'Driscoll, a former vice president and director of policy analysis at Citigroup and a former vice president at the Federal Reserve Bank of Dallas, "At heart, Fannie and Freddie had become classic examples of 'crony capitalism.' The 'cronies' were businessmen and politicians working together to line each other's pockets while claiming to serve the public good. The politicians created the mortgage giants, which then returned some of the profits to the pols—sometimes directly, as campaign funds; sometimes as 'contributions' to favored constituents." [5]

Of course, Fannie and Freddie expect the groups they fund to go to bat for them when they are facing scrutiny. When the Federal Reserve commissioned a study demonstrating that without the federal guarantee they would not be able to take such risks or leverage themselves so much, efforts were made to block it. The Hispanic Congressional Caucus—"at Fannie Mae's urging," according to the *Wall Street Journal*—pressured Fed Chairman Alan Greenspan not to release the study. Fannie had donated $1 million to the Caucus's nonprofit educational foundation not too much earlier and obviously wanted something in return.

Fannie Mae spokeswoman Janice Daue defended the blatant attempt at censorship. "It's an important issue to their constituency. We talk to Congress all the time. It is our job; we're congressionally chartered to talk to Congress." [6]

In short, Fannie and Freddie have created an iron triangle in Washington. There are the well-paid executives who run these enormous enterprises; the government officials and politicians who receive financial rewards for seeing that their special status and federal backing are not jeopardized; and the activist groups that receive lavish funding for their fair housing campaigns. "Crony Capitalism is a good analogy," says Peter Wallison of the American Enterprise Institute. "Investments are being affected by a close relationship between the enterprise and the government, and the money flows back to the government patrons in many forms—political donations, the hiring of government officials, distribution of grants in every congressional district. Everyone is getting paid out of that big trough."[7]

Those who might try to get between the Washington elites and the Fannie Mae trough by questioning the merits and direction that the GSEs have taken soon find themselves in the crosshairs. The agencies and their congressional allies work hard to silence any criticism. Owen Ullmann, a former *BusinessWeek* writer who covered economic policy for *USA Today*, wrote of Fannie Mae, "It will hire key government critics to buy their silence, and it will intimidate lawyers, consultants and financiers who go up against it by pressuring clients of the opponents to withdraw their business." He goes on to quote a congressional source: "Fannie has this grandmotherly image. But they'll castrate you, decapitate you, tie you up and throw you in the Potomac. They're absolutely ruthless."[8] Not what you expect to hear about a government-sponsored agency.

When Congressman Richard Baker, a Republican from Louisiana, tried to rein in Fannie and Freddie back in 2000, he found out how they operated. As a subcommittee chairman he wanted to probe the increasing liabilities at both financial giants. Fannie responded to his efforts at greater transparency by pointedly donating large sums of campaign cash to twenty-one of the subcommittee's twenty-seven other members. Congressman Ken Bentsen, a Democrat from Texas, alone cashed $17,000 worth of checks. The House Banking Committee was also flooded with letters from constituents proclaiming their outrage at the idea of reforming Fannie Mae. The committee discovered, however, that some of the letters had actually come from constituents who were dead. Regardless, there was no interest from most committee members in exploring of these issues further.[9]

When Congressman Jim Leach, a Republican from Iowa, proposed that Fannie Mae should pay a portion of the costs of the savings and loan bailout in 1992, he was "hit with a political steamroller," said the *New York Times*. The company "mobilized much of the nation's housing-finance and real estate industries to attack the proposal as a tax on homeownership. The idea was quickly buried." [10]

When Salomon Brothers protested Fannie Mae's hope of expanding its charter to expand its business, Fannie Mae cut it off from the lucrative underwriting business that Fannie had been giving them. And when *The Economist* magazine ran several articles and an editorial cartoon criticizing the way in which Fannie did business, Fannie withdrew its advertising. [11]

Criticisms of Fannie Mae within the Clinton administration were squelched. When the Treasury Department produced a study that was critical of the government's sponsorship of Fannie and Freddie, Fannie officials went to Treasury Secretary Larry Summers and persuaded him to tone the report down. [12]

The election of George W. Bush in 2000 threatened Fannie and Freddie's nice little arrangement.

It is a myth that the Bush administration did nothing to forestall the looming housing bubble and attendant financial crisis. To the contrary, concerned about rising national debt (brought on in part by the immense economic blow of September 11, 2001, and the costs of launching a war in Afghanistan and later Iraq), the Bush administration tried to rein in Fannie and Freddie. In 2002, the Treasury Department, with support from the Securities and Exchange Commission (SEC) and the Office of Federal Housing Enterprise Oversight (OFHEO), declared in a report that Fannie and Freddie needed to disclose more about their mortgage-backed securities. They had been required to register their common stock with the SEC because they had been chartered by Congress, so they didn't need to file quarterly or annual statements. The Treasury Department wanted them to provide information on their mortgage-backed securities along the same lines as private mortgage lenders had to.

In Congress, Christopher Shays, a moderate Republican from Connecticut, and Congressman Edward Markey, a Democrat from Massachusetts—one of the few on his side of the aisle who would stand up to Fannie and Freddie—pushed legislation that would compel the companies to meet the SEC disclosures mandated for other companies. Facing the pressure of federal action, Fannie and Freddie agreed to these reforms.[13]

However, senior management at Fannie Mae was not happy. Chief Operating Officer Daniel Mudd, who would later become CEO, wrote an e-mail to Franklin Raines in 2002 that would later be released by federal regulators: "We used to, by virtue of our peculiarity, be able to write, or have written, rules that worked for us. We now operate in a world where we will have to be 'normal.' " Mudd made clear in a separate memo how much the Bush administration reform efforts differed from the Clinton years, when Fannie and Freddie had gotten everything they wanted, no questions asked. Mudd wrote in November 2004, "[t]he old political reality was that we always won, we took no prisoners, and we faced little organized political opposition."[14]

White House Chief Economist Gregory Mankiw fired directly at the two government-sponsored entities when he explained the problems with the "perception" that the federal government guaranteed Fannie's debt. "The subsidy creates a source of systemic risk for our financial system," he said bluntly in 2004. The subsidy had allowed the companies "to become gigantic," with a combined debt at the time of more than $2.2 trillion. This "implicit public guarantee," in Mankiw's words, was creating incentives for both firms to take even bigger risks. "The savings and loan crisis of the 1980s illustrates the adverse incentive effects that can rise as a result of government guarantees," he warned.[15]

The response on Capitol Hill among Fannie and Freddie supporters was swift and strong. Congressman Barney Frank fired back, declaring that such criticisms were evidence that the Bush White House was not sufficiently concerned about affordable housing.[16]

To cope with the rising tide of concern in the Bush administration and on Capitol Hill, Fannie and Freddie embarked on an ambitious political plan. In 2001 alone, Freddie Mac held forty fund-raisers for Congressman Michael Oxley, the Republican Chairman of the House Financial Services Committee. They also raised money and held fund-raisers for other committee members, including ranking Democrat Barney Frank. Rahm Emanuel was on the Freddie Mac board at the time and approved of a strategy called "political risk management" that was designed to influence lawmakers by raising money for them and blunting opposition to their plans. Fannie and Freddie sponsored more than eighty fund-raisers for congressional candidates, even though federal law bans corporations from engaging in direct political activity. This blatant violation of the law would lead to a record fine of $3.8 million from the Federal Election Commission.

When Emanuel ran for Congress in 2002, he received more than $25,000 in contributions from Freddie Mac's senior management, more than twice the amount collected by any other candidate for the House or Senate. Once elected, Emanuel took a slot on the Financial Services Committee, where he sat on the subcommittee with direct oversight of Freddie Mac. When the political fund-raising scandal broke, Freddie Mac CEO Leland Brendsel was forced out. Emanuel's subcommittee organized a series of hearings to get to the bottom of the scandal. The hearings lasted more than a year. According to congressional records, Emanuel did not attend a single meeting.[17]

Meanwhile the Bush administration had grown very concerned about how vulnerable the portfolios of these two giants might be. Treasury Secretary John W. Snow, with support from Federal Reserve Chairman Alan Greenspan, urged Congress to limit the size of their portfolios, to "reduce the risk to the federal government and the economy." Their growth had been dramatic—from $136 billion in 1990 to $1.6 trillion in 2003. Much of this growth had come in the form of mortgage-backed securities.

Franklin Raines had pushed Fannie heavily into mortgage-backed securities because of the enormous profits that could be reaped. The spread between the low cost of borrowing and the higher interest rates on mortgage-backed securities—particularly subprime ones—helped management meet their profit goals, enabling them in turn to reap large bonuses and spread more cash around Capitol Hill. This was in no way part of Fannie's original mission, which was to create a liquid secondary mortgage market.[18]

Fed Chairman Alan Greenspan was worried about the fact that the GSEs were seen by other investors as enjoying the full backing of the federal government: "A market system relies on the vigilance of lenders and investors in market transactions to assure themselves of their counterparties' strength," he observed. "However, many counterparties in GSE transactions, when assessing their risk, clearly rely instead on the GSE's perceived special relationship to the government." Investors reasonably assumed that, in the event of a crisis, "policymakers would have little alternative than to have the taxpayers explicitly stand behind the GSE debt."[19]

Because of their special status, these companies could take on huge debt obligations. So Fannie Mae, with a capital base of only $40 billion, could take on debt of $2.8 trillion in mortgage-backed securities and an additional $700 billion in mortgages.[20]

The federal government had created a massive financial distortion in the market. Fannie and Freddie had grown dramatically, and they were not limited by market forces. They did not have to keep much equity on hand to cover their debt. They could borrow more money at lower rates. Treasury Secretary Snow was therefore proposing "to limit the size of the investment portfolios, which consist of mortgages and mortgage-backed securities that are subject to several types of financial risk. If these risks are not managed properly, or if market movements turn dramatically against the GSEs, the government may face two unsatisfactory alternatives: either let the GSE go into default and hope that the financial repercussions can be controlled, or step in and assume payments on the GSE debt at a significant cost to taxpayers."

Greenspan also warned, "Today, the U.S. financial system is highly dependent on the risk-managers at Fannie and Freddie to do everything right. . . . The concentrations of mortgage-backed securities at Fannie and Freddie are well beyond what market forces would normally allow because there are no meaningful limits to the expansion of the GSEs portfolios, which are funded with debt that the market believes to be federally guaranteed."[21] Both Snow and Greenspan were blunt: "The mortgage portfolios of Fannie Mae and Freddie Mac present a risk to the nation's financial system and federal government."

In early 2003, the Senate Banking Committee approved a bill to tighten the regulation of these lenders, thanks to the support of every Republican on the committee. All committee Democrats voted against it. The bill was killed on the Senate floor.

In the House, there was no greater success. In 2003, hearings were convened to promote increased regulation of Fannie and Freddie. "There is a broad agreement that the current regulatory structure for the GSEs is not operating as effectively as it should," explained Congressman Shays. But Committee Democrats were having none of it. Congresswoman Maxine Waters—a reliable supporter of Fannie and Freddie and a strong ally of the housing activists who were now effectively running the show—declared, "Mr. Chairman, we do not have a crisis at Freddie Mac, and in particular at Fannie Mae, under the outstanding leadership of Mr. Frank Raines. Everything in the 1992 act has worked just fine. In fact, the GSEs have exceeded their housing goals. What we need to do today is focus on the regulator, and this must be done in a manner so as not to impede their affordable housing mission, a mission that has seen innovation flourish from desktop underwriting to 100% loans."

Barney Frank—another reliable shill for the affordable housing agenda—was concerned that there was too much focus on the issue of safety and soundness. As he put it during a September 25, 2003, hearing of the House Financial Services Committee: "I do think I do not want the same kind of focus on safety and soundness that we have in OCC [Office of the Comptroller of the Currency] and OTS [Office of Thrift Supervision]. I want to roll the dice a little bit more in this situation towards subsidized housing." With help from Frank, who persisted in his view that everything was working "just fine" at the two mortgage giants, Waters killed reform efforts in the House.[22]

The companies also relied on friends and allies who were paid handsomely to represent their interests. Harold Ickes, the Democrat political strategist and Hillary Clinton ally, was on the Freddie board. Thomas Downey, a former Democrat congressman from New York, was a Fannie lobbyist. Former Senator Alfonse D'Amato, a Republican, was a Fannie consultant. Jamie Gorelick, a deputy attorney general in the Clinton administration, served as vice chairman of Fannie Mae.

And then there were the lobbyists. On the Republican side, the list of Fannie and Freddie lobbyists was a virtual rogues' gallery—Tony Rudy, Edwin Buckham, Kevin Ring, and David H. Safavian—all of whom were linked in some way to the Jack Abramoff scandal.[23] Democrat lobbyists included Maria Echaveste, a top Clinton White House official whose husband, Christopher Edley, would be an Obama confidant, as well as William M. Daley, the commerce secretary in the Clinton administration and brother of the mayor of Chicago. Daley was now an in-house lobbyist. Other Democrats who lobbied on behalf of the GSEs included Ron Klain, a former Gore aide, and Congressman Harold E. Ford.

Fannie and Freddie also took care of their political allies by making generous campaign contributions. Then-Senator Barack Obama received more than $120,000 over the course of three years. (Both Jim Johnson and Franklin Raines would play important roles in Obama's presidential campaign.) Senator Chris Dodd received more than $133,900 in political contributions, making him the number one recipient of campaign funds from Fannie Mae. In return he has been its staunchest defender, insisting again and again despite mounting evidence to the contrary that it was "fundamentally strong" and "in good shape."[24]

Fannie Mae also kept no fewer than twenty high-powered law firms on retainer to help lobby and deal with political problems. Former Iowa Congressman Jim Leach came to the conclusion that its financial clout and ability to pay large fees to members of the Beltway establishment gave Fannie more power with Congress "than all the private banks in the United States put together."[25]

This was a key component of Fannie Mae's political strategy. As the Office of Federal Housing Enterprise Oversight (OFHEO), Fannie's regulator, noted, executives had managed to create the impression that "what is good for Fannie Mae is good for housing and the nation. Senior executives used that image and their political influence to try to ensure that Fannie Mae operated under rules that differed from those that applied to other corporations."[26]

Just how differently Fannie Mae operated became abundantly clear when the OFHEO took a look at the giant's financial records and found something astonishing: Franklin Raines and members of the senior management had been cooking the books. It was bad enough that Fannie was taking on huge amounts of high-risk debt; now came a 211-page report that laid out in detail serious accounting irregularities. The details were clear: Fannie Mae was shifting profits and losses from year to year so it could meet performance targets that in turn triggered management bonuses.

The amounts involved were staggering: an estimated $11 billion had been misreported on financial statements. (A similar investigation of Freddie Mac found that between 2001 and 2004 the company had $5 billion in accounting irregularities; it was later fined $125 million.) To put that into perspective, Enron had overstated its earnings by $567 million and set off a firestorm on Capitol Hill. Congressman Barney Frank had decried the "shameful corruption" at the company, as he should have. But Enron's accounting scandal was but 5 percent of Fannie Mae's.

Soon a cynical joke began circulating in Washington: "What's the difference between Enron and Fannie Mae? The guys at Enron have been convicted." [27]

Franklin Raines denied under oath that accounting problems existed. In Congress the lines were soon drawn. Republicans such as Representative Christopher Shays welcomed the report. "I congratulate OFHEO for finally stepping up to the plate and not being manipulated." He warned that Fannie and Freddie "are going to crash if this Congress doesn't wake up and do something about it." But Democrat Barney Frank, a vocal defender of Fannie Mae who had been harshly critical of Enron, suddenly could not even offer a soft protest about a scandal 11 times as large. "I see nothing here that suggest that safety and soundness are an issue," he said. "It serves us badly to raise safety and soundness issues and not have it there." [28]

Some Democrats went even further. Congressman William Clay, Jr., of Missouri accused OFHEO of conducting a "witch hunt" against Raines because he was black. "This is about the political lynching of Franklin Raines." (Clay is a second-generation black congressman whose father, William Clay, Sr., cofounded the Congressional Black Caucus. In 2007 he vociferously objected to the inclusion of Steve Cohen, a white Democrat who happens to represent a majority-black district in Tennessee.) Curiously, Clay didn't mention that the whistle-blower fired by Fannie after complaining about the accounting procedures, Roger Barnes, was also black. [29]

Democrats on the Financial Services Committee, including Barney Frank, took their cue from Clay and went after the regulator, OFHEO, rather than Fannie Mae. As *American Banker* magazine reported, "Democrats mostly defended the company, charging that the agency's allegations were politically motivated." In a transparent attempt to punish the regulatory agency for doing its job, Barney Frank actually sought a cut in its funding.[30]

The news media also had little interest in the story. A LexisNexis search reveals that the major news outlets blanketed Enron with coverage. CNN alone mentioned the story 1,385 times over an eighteen-month period. The Fannie Mae scandal was mentioned by the entire media—ABC, CBS, NBC, and CNN—just thirty-seven times over the same time period.[31]

The fact that these large financial entities had crooked books raised serious concerns about their financial soundness. They were highly leveraged, increasingly with subprime loans, and their balance sheets were now exposed as murky in the extreme. But Fannie's allies on Capitol Hill were not concerned. Congresswoman Maxine Waters blustered, "We do not have a crisis at Freddie Mac." Congressman Clay remarked, "Markets are not worried about the safety and soundness of Fannie Mae." Franklin Raines went so far as to assert, "These assets [home mortgages] are riskless." It was a claim that anyone with the smallest amount of financial knowledge would find completely laughable.

Yet Congressman Barney Frank still radiated confidence. "I think it is clear that Fannie Mae and Freddie Mac are sufficiently secure so they are in no great danger. . . . Fannie Mae and Freddie Mac do very good work, and they are not endangering the fiscal health of this country." Statements such as these would in the end make Congressman Frank look either foolish or downright incompetent.

Raines was cooking the books because his compensation, and that of other senior executives, was tied in large part to the share price of company stock. Federal regulators were highly critical of

Fannie and Freddie board members, who either paid no attention to the accounting irregularities or purposely looked the other way. As OFHEO noted in its report on Fannie Mae, "The Board failed to exercise the requisite oversight to ensure that the enterprise was fully compliant with applicable law and safety and soundness standards."

Freddie Mac eventually admitted what had been done. As part of the settlement, Freddie paid a fine of $125 million, fired its top management, and restructured its pay system.[32]

Fannie Mae admitted to much the same. In 2004, Raines wisely took early retirement, and two years later he forfeited some stock options and agreed to make a donation to charity. (For "affordable housing," of course.) He also agreed to pay a fine of $2 million to the federal government. But his stock options were already underwater, and Fannie's insurance policy, which covered officers and directors, paid the multimillion-dollar fine. Despite the scandal, Raines, like his predecessor, Johnson, would go on to serve as a confidant of Barack Obama.[33]

But the problems of Fannie and Freddie did not go away. The march to financial oblivion continued and was enabled by supporters on the Hill. In 2007, as the housing bubble burst, Senator Schumer of New York and Congressman Frank declared that Fannie and Freddie could solve the growing crisis—by buying up even more subprime mortgages! Schumer's "Protecting Access to Safe Mortgages" bill would have allowed Fannie Mae to increase its portfolio by another $145 billion in new mortgages. When the Bush administration and Federal Reserve Chairman Ben Bernanke opposed the idea, Frank sharply criticized them for failing to support affordable housing.[34]

As late as July 2008, Senate Banking Committee Chairman Chris Dodd was explaining that Fannie and Freddie were "fundamentally strong" and that the financial concerns were not real. "This is not time to be panicking about this. These are viable, strong institutions."[35]

Days later, on July 30, President George W. Bush signed a bill to bail out Fannie Mae and Freddie Mac with $25 billion of taxpayers' money. But it is only the beginning. Some economists believe that taxpayers may be on the hook for more than $1 trillion in bad Fannie and Freddie debt when all is said and done.

Those who created the mess, of course, have been handsomely rewarded. Daniel Mudd, the head of Fannie Mae, earned $80 million in his eight years at the company and received a nice retirement package and deferred compensation. He has pocketed $12.4 million since becoming CEO in 2004. Richard Syron, the departing head of Freddie Mac, pocketed $14.1 million, along with the $17.1 million in pay and stock options he had received in 2003. Syron, of course, was the Fed official who in 1992 had authorized the famous study on alleged racism in lending.

Vernon Smith, a Nobel laureate in economics, says the federal government "set the stage for housing bubbles by creating those implicitly taxpayer-backed agencies, Fannie Mae and Freddie Mac as lenders of last resort."[36] In other words, far from being the results of runaway capitalism, the housing bubble and the financial crisis that followed are the results of runaway government. Through Fannie and Freddie, the federal government was collaborating with the subprime lenders, encouraging their growth as a rational response to federally dictated changes in the market.

As we shall see in what follows, the Clinton administration was simultaneously doing much the same thing on Wall Street, using the power of government to foster a new kind of state capitalism in partnership with the big financial firms. The results were no happier than they were with Fannie and Freddie's interventions in the housing market.

DO-GOOD CAPITALISTS

Bill Clinton's Seduction of Wall Street and the Birth of the Bailout Culture

Capitalism without risk is like religion without sin.

—ALLAN MELTZER, CARNEGIE MELLON UNIVERSITY

Goldman Sachs. Citigroup. Salomon Brothers. J.P. Morgan. The names conjure up images of naked capitalism, titans of finance that succeed or fail on the basis of risk. Taking their wins and losses, they are not only bastions of wealth but temples of the free market.

So when the financial crisis exploded in 2008 and these firms were drained of cash from being overleveraged and indulging in extraordinary risks, the culprits were seemingly easy to identify: greedy capitalists whose appetite for profit had nearly brought down the whole house of cards. What's more, they had been encouraged in their ravenous appetites by a system of naked capitalism that promoted such excess and resisted government oversight. Federal authorities should have stepped in much sooner to protect us from these profit-crazy vandals, but conservatives in Congress and the White House saw to it that restraints were undermined, creating a volatile, unregulated economic system just waiting to implode.

That is a nice narrative as far as it goes. And it is true that Wall Street used to operate at the tip of the capitalist wave. Back in J. P. Morgan's day, Wall Street was deeply involved in the industrialization of America, financing railroads and telegraphs, shipyards and factories. Speculation was rampant and unregulated, and fortunes were made and lost overnight, which is why investing in stocks was always regarded as a risky proposition, not for the faint of heart.

But Wall Street today is a very far cry from the arena of freewheeling capitalism most people recall from their history books. Indeed it hardly deserves to be called a market-based system at all. A properly functioning market, after all, regulates foolish and risky behavior by ensuring that extreme risks fail more often than they succeed. But things don't work that way on Wall Street anymore. And it was in the Clinton years—not those of Reagan or Bush—that the culture of Wall Street began to change. The driving cause? An unprecedented partnership between the financial industry and the federal government.

Does the idea surprise you? If so, consider this simple question: how often have these supposed temples of free-market capitalism been bailed out by the federal government in the past twenty years? Many will assume that the bailouts of 2008–2009 were an anomaly, the first time these proudly independent firms were ever rescued by taxpayers. But the reality is that *all* of these supposed paragons of free-market capitalism have been saved at least once from the effects of their enormously risky and often stupid financial decisions. Some, such as Citigroup and Goldman Sachs, have gone with their hand out to the federal government four times in the past fifteen years.

Unadulterated capitalism is all about risk and reward. Take big but educated risks, and you may profit enormously. Take the wrong risks, and you will end up losing your shirt (and your investors' as well). But what happens if you make reckless financial decisions and are repeatedly bailed out by taxpayers? You don't need a Harvard MBA to figure that one out.

When people think of the 1990s, they think of the dot-com tech bubble. But they tend to overlook a series of other bubbles that were even more important and that blazed the path for the subprime mortgage meltdown. At the heart of these massive economic distortions was the mercurial Bill Clinton, the man whose impulsivity, grandiosity, and ethical corner cutting conferred the name "Clintonian" on the decade in which he was president.

Previous generations of Democrats had been economic populists, aligning themselves with "the little guy" against the "money power" of Wall Street. When Clinton decided to run for president in 1992, he struck this familiar anti-Wall Street tone, declaring, "Never again should Washington reward those who speculate in paper, instead of those who put people first." ("Putting people first" had been the Clinton campaign slogan.)

Yet in 1991, as he was gearing up for his presidential run, he forged a close alliance with a powerful Wall Street insider, Robert Rubin, the cochairman of the investment giant Goldman Sachs, who had been a very successful fund-raiser for the Democratic Party since 1982. A dedicated liberal, Rubin liked what he saw in the young Arkansas governor and brought him to New York for a series of meetings and fund-raising events with other Wall Street heavy hitters. Indeed, investment bankers would dominate financial giving to the Clinton campaign. Daniel Gross captured the new attitude perfectly when he wrote in his book *Bull Run*, "It's more hip to be a Democrat if you're loaded than it was in the 1980s. And it's more hip to be loaded if you're a Democrat."

When Clinton won in 1992, he asked Rubin to run the newly formed National Economic Council. When Lloyd Bentsen retired as Treasury secretary in January 1995, Rubin would replace him.

Rubin was emblematic of the arrival en masse of baby boomers in the Wall Street executive suite. Once dominated by the Depression generation, which had grown up averse to risk and valuing thrift, the boomers arrived on Wall Street with the same proclivities for excess, self-absorption, and entitlement they had displayed since their emergence in the sixties. To the boomers, who had grown up in the midst of American abundance and prosperity, poverty and want were just a rumor. "Fearing something you've never faced is difficult," explained the economist and money manager Gary Shilling about the new generation. They were also smart, well educated, and highly confident in their abilities. All this translated into an attitude of boundless (some might say reckless) optimism. *BusinessWeek* announced that the World War II generation and the Depression generation were now being replaced by "the Bull Market generation." When Alan Greenspan worried at the height of the nineties tech bubble about "irrational exuberance," he was quickly shouted down by high-flying boomers on Wall Street who spoke of a "new age" economy that could only go up.[1]

In this regard they were very much like Clinton himself, and as a group they displayed much the same mix of narcissistic self-indulgence and sixties-era liberal nostalgia that characterized the Clinton Democrats in Washington. Together, these two groups—the Washington and Wall Street branches of the emerging boomer overclass—forged a new form of liberal state capitalism.

The boomers began to enter Wall Street in the late 1960s and early '70s and rose quickly (Rubin started at Goldman Sachs in 1966). By 1994, three of the top five Wall Street firms—Goldman, Salomon, and Lehman Brothers—were all headed by boomers such as Jon Corzine of Goldman Sachs. If in the sixties they were going to bring about world peace and overthrow repressive social values, they now set out to remake Wall Street in their own image. Three Wall Street newcomers founded an "anti-Establishment" charity called the Robin Hood

Foundation and organized "guerrilla benefits" replete with Hollywood celebrities. They became the hottest tickets on Wall Street.[2]

"This is a new generation," explained Roy Smith, a professor of finance at New York University and a former general partner at Goldman. "The markets have gone through such rapid change in the past ten years in size and complexity. Many of Wall Street's top executives have experiences that are no longer relevant. That is why guys like Corzine are rising to the top. When an industry goes through changes, youth bails you out."

Jon Corzine is fairly representative of this generation. Corzine was born in 1947 on a small family farm in Illinois and began his banking career in the Midwest. In 1975, he moved to New Jersey to work for Goldman Sachs and over twenty years rose from bond trader to CEO. By the time he was forced out in 1999, he had become extremely rich—rich enough to finance successful runs for senator and then governor of New Jersey. (Corzine reportedly spent $80 million of his own money; under the circumstances it was hard not to conclude that he had "bought" the office as though it were a house in the Hamptons.) But even as he rose to the top of the financial and political establishments, like many of his generation, Corzine never abandoned his youthful roots in the counterculture. Thus (to take a small illustration) he habitually said "Peace" instead of "Good-bye." "I was a child of the '60s, and [that] is probably the last telltale sign, other than my beard."[3] A reliable liberal Democrat, he has consistently pushed for affirmative action, same-sex marriage, gun control, and universal health care.

The boomers had other common characteristics, among them a penchant for sophisticated new management tools and precise mathematical models that could transcend the business cycle. It was thought that these models, developed by specialists known as "quants," could fix inefficiencies and imponderables in the financial system. It was all highly reminiscent of the whiz kids who took over the Department of Defense under Robert McNamara. The investment banking guru Felix Rohatyn warned about these math whizzes and their "financial hydrogen bombs." The legendary investor Warren Buffett likewise declared that America should "beware of geeks bearing formulas."

These comments, which represented the traditional view of economics and investment strategy, also echoed the warnings of the Austrian economist Ludwig von Mises, who in his book *Human Action* denounced "those economists who want to substitute 'quantitative economics' for what they call 'qualitative economics.' " The problem, according to Mises, was not the lack of "technical methods for the establishment of measure." Rather, it was that human behavior does not fit neatly into mathematical formulas. Human actions can't be predicted with certainty.[4] But this was all decidedly old school.

Perhaps most of all, the new generation of Wall Street titans believed in their ability to make the world a better place through financial wizardry, which is to say through the sheer manipulation of money. Indeed, people such as Robert Rubin saw finance as leading the way to a new postindustrial era, one "in which services, especially the lucrative financial ones, would replace manufacturing."[5] There was grandiose talk of how financial markets and digital technology would combine to produce a "long boom" that might go on for forty years. In Silicon Valley, "mind share" was said to be more important than "market share." In the new globalized economy, manufacturing jobs would migrate to developing countries, while in the United States, with its emerging "knowledge economy," education would count for much more than manual skills. Glen Meakem, an innovative thinker and billionaire, likened the move to the "new economy" as the equivalent of going to Woodstock for a generation who thought they had rewritten the rules of capitalism, the way Jimi Hendrix rewrote the rules of the electric guitar.[6]

This giddy confidence in the promise of technocracy proved to be highly infectious. Thomas Friedman of the *New York Times* claimed that "international finance has turned the world into a parliamentary system." The financial journalist Daniel Gross wrote a well-received book (mentioned above) whose first chapter was called "The Democratization of Money" (a title curiously reminiscent of the rhetoric of Alinskyite housing activists). Gross hailed Wall Street's shift from a bastion of Republicanism to one of the pillars of the Democratic Party and argued that "arrogant capital," the top-down kind practiced by conservatives, was being replaced by "humble capital," which happened to be advocated by liberals.[7]

Needless to say, this assessment is utterly laughable. The new liberal approach to capitalism was anything but "humble." For the new breed of Clintonian Democrats—on Wall Street and elsewhere—it was blindingly obvious that smart liberals like themselves should run the world. If they happened to get rich in the process—well, there was nothing wrong with that.

Of course, by moving closer to Wall Street, Clinton was moving away from the traditional Democrat constituency of organized labor. He and his acolytes, members of a new class of liberal technocrats, embraced the rhetoric of globalization with its happy talk about the promise of emerging markets because it reflected the interests of his clients on Wall Street, who were reaping huge profits from the often volatile process of forging a global capitalist market. NAFTA—heavily promoted by Clinton, bitterly opposed by domestic labor unions—was an unmistakable sign of this powerful new alliance.

Nor for their part were the new masters of the universe averse to government involvement in the market. To the contrary, they praised such intervention in the name of social justice. Thus, in the past, the titans of Wall Street had condemned the Community Reinvestment Act as a form of credit allocation—a threat to sound banking practices. Now Robert Rubin defended it in the pages of the *New York Times* under the headline "Don't Let Banks Turn Their Backs on the

Poor."[8] In short, government was no longer seen as a threat but as a partner. It was the essence of do-good capitalism: the merging of sixties social values with the rewards of the profit system.

Unlike earlier generations, who valued the free-market system (despite its flaws) for its ability to generate wealth and maximize personal liberty, the liberal baby boomers—born to affluence, burdened by guilt—saw the capitalist system as inherently flawed and unfair. More to the point, they saw it as a system that could, and should, be manipulated for "progressive" social purposes. And as the liberal boomers rose to positions of power in the 1980s and '90s, they increasingly sought to harness the engine of capitalism to their vision of a good society. Rather than engaging in financial transactions purely for the sake of profit, do-good capitalists seek to use capitalism to advance their "transformative" social goals. The chief buzzwords of this enlightened form of capitalism are the now-fashionable notions of socially responsible investing and corporate citizenship. These benign-sounding terms have turned out to be powerful tools of "soft" coercion, resetting the moral compass of the American corporate elite in a decidedly leftward direction and lining CEOs up like iron filings.

Traditionally sympathetic to Republicanism, the new masters of finance began to skew heavily left. As Kevin Phillips notes, "In deference to their multiple Democratic coalition mates, they donated to the NAACP; joined the boards of environmental groups; embraced technology, education, free trade, and globalization; and worried about the growing international gap between the rich and the poor as well as the gap in the United States."[9] Thus Sanford "Sandy" Weill, the cochairman of Citigroup, was a big supporter of Jesse Jackson and cochair of his Wall Street Project. One of Jackson's biggest backers was Kevin Ingram, who ran the mortgage-backed securities desk at Goldman and was later at Deutsche Bank. A protégé of Robert Rubin, Ingram donated $50,000 to Jackson's Citizenship Education Fund, his chief tax-exempt group, in 2000. (Ingram was later convicted on federal money-laundering charges related to a Pakistan arms deal.) [10]

Meanwhile, charitable giving by the big Wall Street investment houses shifted its focus from strengthening the free-market system to supporting activist causes and groups, some of which were busily engaged in undermining the roots of the financial system and propelling us toward an economic crisis. Whereas they had once resisted the likes of the militantly anticapitalist ACORN and other groups that tried to shake them down, the big firms now embraced them— or at the very least hoped to befriend them with lavish contributions. JPMorgan Chase gave $5 million to ACORN, Bank of America $1.4 million, and U.S. Bancorp more than $700,000. (J. P. Morgan himself must have been rolling in his grave!) Citigroup Foundation also gave large grants to Jesse Jackson's RainbowPUSH Coalition. Hardly a dime went to free-market think tanks. In a very real sense, many of the large firms on Wall Street were following the pattern established decades earlier, when Wall Street financiers had underwritten the efforts of Saul Alinsky. Only this time they weren't doing it under duress. [11]

Clinton and Wall Street would enjoy a wonderful eight-year marriage. As with the wealthy liberals of Hollywood—the third pillar of the emerging boomer overclass—it had been love at first sight. Wall Streeters were impressed with Clinton's charm and wide-ranging intellect. Clinton in turn was impressed by the wealth and power of the world of finance. Lavish parties and fund-raisers in the Hamptons and on Wall Street were hallmarks of the Clinton years.

Meanwhile, unnoticed at the time, Clinton's Washington and Rubin's big Wall Street firms formed a corrupt relationship in which Washington would seek to minimize the impact of market forces and Wall Street would provide campaign funds and business opportunities for Clinton and his Democratic allies.

Robert Rubin was a believer in the so-called virtuous cycle that linked the health of the securities markets to that of the national economy. Clinton came to embrace that view with great enthusiasm. And in an era of post-Cold War peace and prosperity, U.S. foreign policy was increasingly driven by economic interests. As Steve Fraser attests in his history of Wall Street, "both the domestic and foreign policy of the Clinton administration pursued the objectives coveted by Wall Street's biggest interests."[12] Instead of the laissez-faire attitudes of Ronald Reagan and George H. W. Bush, who embraced the notion of market discipline, the Clinton administration was willing to provide a safety net, one that would stymie market forces through government-big business collaboration. Wall Street, which had traditionally been suspicious of the federal government, would come during the Clinton years to see it as a friend that would protect its profits and rescued it from the consequences of its own risky behavior.

After the collapse of communism in 1989, global markets opened up to capitalism and a Wild West atmosphere prevailed in the investment world. With the knowledge and encouragement of government, the investor class plunged headlong into an era of global ferment. This Clintonian approach to economics—relentlessly upbeat and optimistic, heedless of consequence, really a kind of economic promiscuity—set off a chain of economic bubbles all over the world. It is well worth looking back at them to see how they set the stage for the domestic crisis to come in the following decade.

In the early 1990s, one of the hottest investments on Wall Street was Mexican government bonds. It was a business with which Rubin himself was intimately familiar. During the 1980s, Rubin and Goldman had advised Carlos Salinas, who was then the Mexican secretary of planning and budget, on how to use hedge funds to protect the peso from currency changes. Salinas (who would later become Mexico's president) in turn gave Goldman Sachs a contract to underwrite the $2.3 billion international public offering of the national telephone company, Telmex.

This naturally led to more business contacts in Mexico. In his financial disclosure upon joining the Clinton administration as head of the National Economic Council, Rubin listed six Mexican clients, including the Mexican government, the finance ministry, the central bank, Cemex, a large cement company, Telmex, and DESC Sociedad de Fomento, an industrial and manufacturing conglomerate.[13]

Buying Mexican bonds was profitable for firms like Goldman and other large institutional investors because they were using "arbitrage," that is, borrowing money in New York at 5 percent and then buying Mexican government bonds that paid 12 percent. It was a good but risky deal. These firms were counting on the Mexican government, which was rife with corruption, waste, and fraud, to fight the drug cartels, face down labor unrest, battle an unstable oil market (Mexico is a major exporter), and struggle through a financial crisis that was already raising questions about its financial soundness.

In retrospect, of course, one can see how foolish it was to invest large sums of money in a far-off market whose local conditions could not be grasped from an office on Wall Street. And this should have been apparent to sober investors at the time.

By mid-1994, the prospect of the Mexican government making good on these bonds was looking pretty dicey. Larry Summers, a Harvard-trained economist who had joined the administration as undersecretary of the Treasury for international affairs, went to Rubin and explained that the Mexican government was $25 billion or more short and couldn't make the payments. Goldman Sachs, Citibank, and other big financial houses deeply invested in Mexico were suddenly facing enormous losses.[14]

Rubin was under no illusion as to who was at fault. He laid the blame for the crisis squarely at the feet of the big institutional investors on Wall Street who had taken foolish risks. Ninety percent of the

Mexican bonds were held in New York, and Goldman Sachs itself had been the largest single underwriter of Mexican equities and bonds from 1992 to 1994. Rubin's old firm reportedly had $5 billion of its clients' own money tied up in Mexico, more than any other investment bank in America.[15]

As Rubin later put it in his memoirs, "The Mexican crisis is usually viewed as a failure of Mexican policy. But it was, crucially, also a failure of discipline on the part of creditors and investors. . . . Lured by the prospect of high returns, investors and creditors hadn't given sufficient consideration to the risk involved in lending to Mexico."[16]

If you are going to chase high returns by taking big risks, you need to be prepared to face the consequences when your investments go south. If you don't, you are likely to be just as foolish (if not more so) in the future. In the investment world this is known as "moral hazard." Greed is rampant on Wall Street; people are there, after all, to make money. But no one—especially the greedy—likes to lose money, so the threat of losses produces fiscal discipline.

Bill Clinton was not a student of moral hazard, and with large investment houses facing huge losses, he took an unprecedented step: almost immediately he began pushing for a federal bailout. It was a fateful step and one that turned into a habit that, once formed, would prove impossible to break.

There had been government bailouts before. The savings and loan (S&L) debacle of the late 1980s had cost taxpayers an estimated $160 billion. In 1984, the Reagan administration, while "holding its nose," in the words of one observer, had helped in the bailout of Illinois Continental, a midwestern bank that had been caught in a cash crunch. But in those instances, deposits at the banks and S&Ls had been guaranteed by the federal government through the Federal Deposit Insurance Corporation (FDIC). The federal government and taxpayers were already on the hook. The government was required by law to protect banks. In the case of Mexican bonds, however, we were talking largely about investment bankers and brokerage firms who had no such guarantee. This represented a dramatic change in the bailout game and a new stage of American capitalism.

Letting the Mexican government default was not even considered as an option. Rubin, Summers, and then-Treasury Secretary Lloyd Bentsen all went to Congress for billions of dollars in taxpayer money. They spoke in apocalyptic terms of the results if the money were not made available. If Mexico lost the confidence of international investors, they opined, "the entire world was now at risk." If Mexico were allowed to default, large portions of the world's financial infrastructure would go down with it. Rubin would later say that if there had been no rescue, a *global meltdown* would have occurred.

Congressional hearings were quickly organized. The Senate Foreign Relations Committee heard from *Forbes* publisher Steve Forbes, the economist Larry Kudlow, and Bill Seidman, the former head of the FDIC. All three testified that the bailout was a bad idea that would encourage further risky behavior by investment bankers. A Mexican default would cause some discomfort, but government intervention should be avoided. After the hearings, Larry Summers remarked that "great" could now be added to the word "depression" if a bailout failed to go through. For its part, the Federal Reserve concluded that even with the worst-case scenario of a Mexican collapse, the effects on the U.S. economy would be relatively minor. Economic growth would be reduced by only one-half of one percent.[17]

Despite the heated rhetoric and apocalyptic language, conservative Republicans in Congress were not buying the Clinton line. The debate on the floor of the House quickly became personal. Congressman Steve Stockman, a Republican from Texas, charged that it was a bailout for Goldman Sachs, which had underwritten the privatization of some of Mexico's nationalized industries. Stockman argued that Rubin's firm might now face liability from investors who had lost money there. Goldman had indeed been a prime underwriter of Mexican bonds both before and immediately after the passage of NAFTA, as well as securities in connection with the privatization of the Mexican national phone company.[18]

Rubin rejected the notion that the bailout was some insider deal. It was, he claimed, for the benefit of millions of small investors who had given their money to Goldman and Citigroup to manage. "This isn't the old days where there are three rich market manipulators in a back room threatening to pull the pin. Now it's everyone." Rubin explained that he was working to protect "several million investors, participants in pension and mutual funds that have rushed into international markets."[19]

But many commentators were skeptical. As Michael Prowse, a columnist for the *Financial Times*, put it, "The fundamental question raised by the Mexican crisis is whether global investors are going to be treated as adults or children. Are they to be responsible or not?"[20]

With conservatives in Congress rallying against a taxpayer bailout, the Clinton administration decided to go around them. On the night of January 30, 1995, the Treasury Department announced it was providing a $20 billion line of credit to the Mexican government and pushed the IMF to provide an additional $17.8 billion. The funds came from the Exchange Stabilization Fund, which had been created in 1934 when the United States went off the gold standard and was to be used to minimize currency fluctuations. Rubin later called it "the largest nonmilitary international commitment by the U.S. government since the Marshall Plan." To secure the deal, he sent Summers to Mexico to make sure that the Mexican government would offer high interest rates to foreign investors.

The media was euphoric, seeing it as a stroke of financial and diplomatic brilliance. The Mexican bailout was hailed by Tom Friedman of the *New York Times* as "the most important foreign policy decision of the Clinton Presidency." Larry Summers, an architect of the bailout, would later tell Congress that this unusual government intervention in the financial markets had been fully justified. "The right lessons are being learned." [21] Or were they?

This was a dramatic first step in the rise of state capitalism on Wall Street. Bill Clinton clearly understood that by bailing out Goldman Sachs, Citigroup, and the others, and seeing to it that they all profited handsomely from their foolishness, he might be encouraging further risky behavior. In his postpresidential memoir he reflected, "By giving the money to Mexico to repay wealthy investors for unwise decisions, we might create an expectation that such decisions were risk free." [22] But as a man who thrived on risk in both his political and personal life, worrying about consequences seemed to take a backseat to the thrill of the risk itself.

Even more concerned was Hans Tietmeyer, the president of the Bundesbank, Germany's central bank. Tietmeyer, who had once studied to be a Catholic priest, had grown up in a small village near the Dutch border, which (he said) "strongly influenced me with its Catholic church culture, mixed with a dose of Prussian discipline." In other words, Bill Clinton he was not. He was well known for giving heavy doses of tonic to investment bankers who had lost their financial sobriety. Tietmayer was outraged by Clinton's profligate Wall Street bailout. Germany, like Great Britain, had not been consulted. But, more important, he fixated on the issue of moral hazard. Goldman and the others, he thought, should at least suffer "a haircut" for their irresponsible and risky behavior.

As Paul Bluestein of the *Washington Post* recounted in his book *The Chastening*, "It was a point the Clinton administration could not deny. American investors and brokerage firms had bought tens of billions of dollars worth of short-term Mexican government bonds, called tesobonos, and the rescue package was providing Mexico with enough dollars to ensure that it could avoid defaulting on any of those bonds." In short, at taxpayers' expense, U.S. investment houses were walking off unblemished and unchastened, with extra money in their pockets.[23]

In his memoir, Clinton abandons the apocalyptic language used to justify the bailout and instead says that it was important to assuage America's guilt about its past. By failing to support the bailout, "we would be sending a terrible signal of selfishness and shortsightedness throughout Latin America. There was a long history of Latin American resentment of America as arrogant and insensitive to their interests and problems." In short, he claims, he engineered the bailout for the benefit of the Mexican people and for the sake of better U.S.-Latin American relations.[24]

But of course, the bailout benefited investment houses in New York more than it did the Mexican people. As Mexican Foreign Minister Jorge Castañeda noted, "The fund managers and stock brokers of New York knew what they were doing when they invested in Mexican stocks and received colossal returns; they were taking a risk. Now, thanks to the package, the cost of that risk (which produced lavish returns for two or three years) has been transferred completely to the Mexican taxpayer."[25]

The bailout not only benefited the big investment houses on Wall Street, it also benefited Robert Rubin personally. After he left government, Rubin would return to Mexico as a senior advisor to the financial giant Citigroup, less than five years after he bailed it out of its Mexican problem. While at Citigroup, he would negotiate the firm's $12.5 billion acquisition of Mexico's leading bank, Banamex. He remained on its board, pulling down $115 million over the course of nine years.

It's important here to draw a critical distinction: there is an enormous difference between being probusiness and pro–free market. The former position, which the Clinton administration embraced, is concerned primarily with the health of large businesses, in this case the big financial houses. Being pro–free market means being concerned with the health of the capitalist system as a whole and its underlying principles of profit and loss, risk and reward. The Mexican bond debacle was not the first time U.S. companies had pushed into Latin America and invested large sums without fully considering and understanding the risks. But it was the first time the federal government had used taxpayers' money to protect them from the consequences.

In 1981, when Mexico defaulted on tens of billions of dollars in debt, Brazil, Argentina, and other countries in Latin America followed suit. For eight years the Reagan administration did nothing, leaving the banks to sort it out themselves. In 1989, the Bush administration agreed to intervene but was not about to reward the banks for their foolish behavior. Instead, Treasury Secretary Nicholas Brady pushed what became known as Brady bonds. The government brokered a deal whereby U.S. banks would accept 50 cents on the dollar. There was no taxpayer bailout; banks took it on the nose for their irresponsible investments, which lost 50 percent of their value.

This was the manner in which these matters had been handled in the past: those who made foolish investments paid the consequences. But now we were entering the brave new world of risk-free Clintonian state capitalism.[26]

A new consensus was emerging among liberals in Washington. Barry Bosworth served in the Johnson administration and later as the director of Jimmy Carter's Council on Wage and Price Stability. Now at the Brookings Institution, he explained in 1998, "The idea that governments can just stand aside does not work; the costs are too great. The new view is that you cannot tell private parties you won't bail them out and they have to suffer the consequences of the marketplace."[27] This was a brave new phase of capitalism, of public-private partnerships, buoyed by an economy that would always go up.

Soon there was trouble brewing in Asia, where several countries—particularly Thailand, Indonesia, and South Korea—were having difficulty making good on billions of dollars in loans that they had taken out from largely the same firms that had been in Mexico. The situation was different in one respect: this time, most of the loans had been made not to governments but to private Asian banks, investors, and real estate speculators. U.S. investment houses, as well as Asian and European banks, had poured billions into speculative investments, often with little knowledge of what they were actually buying. Now—surprise, surprise!—many of those loans were going bad. The result was a series of currency devaluations, and international investors from Asia, Europe, and the United States were about to take it on the chin for foolishly lending money without doing much due diligence.

The crisis was made worse because the devaluations created chaos in the world of financial derivatives. Once called synthetic securities, derivatives are securities that are, as the name implies, derived from something else. A derivative can be an option to buy a stock or something more complex, such as multiple currencies. Derivatives made the crisis worse because they made it easy to place high-risk bets and leverage money—and American investment houses had made huge sums from selling derivatives in Asia. Indeed, in countries such as Thailand and Indonesia, they made more money selling derivatives than they did making loans. Bankers Trust, J.P. Morgan, and Chase were all neck-deep in selling derivatives in Asia. Now they were facing huge financial losses.[28]

Consider the case of South Korea, a prosperous country but one with a financial sector that was rampant with cronyism, corruption, and a complete lack of transparency. Large U.S. banks had loaned billions of dollars to South Korean companies without paying much attention to what they were doing. Chase Manhattan was in for $5.4 billion, J.P. Morgan for $3.4 billion, Citicorp for $2.8 billion. Many of these investors had had no clue as to what they were actually investing in and no sense of risk about what they were doing.

Robert Rubin, who was now Treasury secretary, would later recall, "I remember, at the time of the South Korea crisis, being struck in discussion with a prominent New York banker by how little he and his company knew about a country to which they had extended a considerable amount of credit. . . . Though the basic hazard of investing in countries with major economic and political problems should have been obvious, the prevailing mentality was to downplay or ignore those risks in the 'reach for yield.' "[29]

Rubin admitted that U.S. investment banks had not done their homework. As he laid it out, the problem in Asian countries was "crony capitalism." The close links among governments, banks, and corporations led to "fundamentally unsound investments by corporations funded by unsound lending from banks." The problem was compounded by foreign investors who "injected an extraordinary amount of capital into these flawed systems without due weighting to the risks involved."[30] Yet again, the investor class had treated the global market as though it were a unified field, operating everywhere by the same rules, regardless of local conditions. Such unified fields exist only in the domain of theoretical physics.

Once more, the Clinton administration decided almost immediately to step in with taxpayers' money. By doing so they made sure not only that the banks and investment houses were protected but that they made a nice return on their investments. This is the essence of state capitalism: the profits go to the financial firms, the losses are covered by taxpayers.

As the crisis began to unfold, the Fed chairman, Alan Greenspan, asked Treasury to look into the possible consequences of a South Korean collapse. How bad would it get if we simply let the investments default? Greenspan, a onetime disciple of Ayn Rand and Milton Friedman, found government bailouts distasteful and felt it was best to let the parties involved settle the matter themselves. Indeed, as Bluestein recounts, "the failure by a major country to pay its obligations might be the best outcome for the financial system, he thought, because if lenders paid the price for having made irresponsible decisions, the moral-hazard problem would be obliterated and justice would be served." But Rubin and Summers would not even let the idea be studied. Failure and lessons learned by reasserting market discipline were not an acceptable outcome.[31]

The Republicans who ran Congress at the time were not eager to cough up tens of billions of dollars in taxpayers' money to bail out Wall Street. So the Clinton administration, working through the IMF, began throwing money at the problem independently. It tapped Treasury Department funds, as it had done during the Mexico crisis. But throwing cash at the problem did little to stem the tide. In August 1997, Thailand's currency, the baht, fell after an IMF-led rescue totaling $17 billion was announced. In Indonesia, the results were no better. A $33 billion package of loans assembled by the IMF at the end of October 1997 generated only a brief rally in the Indonesian rupiah, which then started falling again.[32]

These spasmodic Clinton bailouts did not go unopposed. There was plenty of resistance on Capitol Hill, despite a full-court press by the administration and its allies on Wall Street. Rubin and Summers warned that failing to provide the bailout money would be like "cutting off the water to the fire department when the city is burning." Investment houses and large banks hired lobbyists to secure $18 billion in taxpayers' money for the Asia bailouts. Congressman Richard Gephardt, the House Democratic leader, promised to go to the mat.[33]

But there was simply too much resistance. As one history records, "The administration encountered particular difficulty with the GOP conservatives opposing the bill, since among their number were influential members such as the house Majority Leader Dick Armey of Texas; the majority whip, Tom DeLay, also from Texas; and James Saxton of New Jersey, chairman of the Joint Economic committee.

They derided the Treasury's contentions about the urgency of getting cash to the IMF" [34] Congressman Spencer Bachus reflected the attitude of many conservative opponents when he went to the House floor and explained that since the government expected welfare mothers and small-business executives to take care of themselves, the same message "ought to also apply to rich Greenwich, Connecticut investors who are multimillionaires." [35]

Another voice opposing the bailout was that of Bernie Sanders, the Vermont independent and professed socialist who joined forces with Congressman Ron Paul of Texas, an extreme libertarian, in what must be one of the oddest alliances in recent American politics.

Clinton would have to get creative to bail out Wall Street again. In late December, another rescue was staged for South Korea. The administration immediately went to work at the IMF, going around Congress as it had done in the Mexican case, and promised that the U.S. taxpayer would back a bailout plan. David Lipton, the undersecretary of Treasury for international affairs, was dispatched to South Korea and stayed in a hotel with the IMF mission to explain what the United States wanted in the rescue plan: namely, that American investors would get their money out of it. (Lipton would go on to serve as a senior executive at Citigroup and an adviser to Barack Obama.) South Korean bankers were literally being driven to suicide, but U.S. investment banks such as Citigroup would get a healthy return on their investment when all was said and done. Free-market purists, such as Nobel Laureate Milton Friedman, were appalled: "The effort is hurting the countries they are lending to, and benefiting the foreigners who lent to them." Similar deals were struck to deal with the crises in Indonesia and Thailand. [36]

The big Wall Street firms and banks, on the other hand, were *ecstatic*. As Bluestein recounted in his masterful history of the financial crises and bailouts of the 1990s, "In a sense, the international banks got away with murder. They had foolishly injected billions of dollars of short-term loans into a country with a shaky financial system, yet they were suffering no losses." Eventually the banks reached an agreement with the Korean government to reschedule $22 billion of short-term interbank debt owed by Korean banks. The banks in return received bonds fully guaranteed by the Korean government. These bonds were very profitable investments, paying 2.5 percent over London Interbank Offered Rate (LIBOR), for two-year bonds and 2.75 percent over LIBOR for three-year bonds.[37]

Not surprisingly, news of the bailout sent the stock prices of the major financial institutions straight up. They had escaped yet again, unscathed and with extra money in their pockets, courtesy of the American taxpayer.[38]

The Asian bailouts were seen as a masterstroke by the mainstream media, which, as is usually the case, focused only on the immediate crisis rather than the larger picture. *Time* magazine, under the headline "The Committee to Save the World," boasted a cover photo of Rubin, Summers, and Greenspan with their arms folded, looking cool and confident as they surveyed the world. Rubin liked the cover so much that he had it framed and mounted on the wall behind his desk at Citigroup after he left Washington.

But as Paul Blustein of the *Washington Post* recounts, "the false impression that the international economy was in the hands of masterminds coolly dispensing remedies carefully calibrated to tame the savage beast of global financial markets" was a total myth. "The reality is that as markets were sinking and defaults looming, the guardians of global financial stability were often scrambling, floundering, improvising, and striking messy compromises."[39]

For those paying attention, the implications for the health of the U.S. financial system were frightening. In the words of one observer, with the Clinton-sponsored bailouts, "it was easier for the greedy and ignorant to gamble with other people's money. If they gambled and lost, if they were important enough, they had to be rescued."[40] The hedge fund manager David Tepper understood what was going on. The big banks and investment houses were operating under the impression that "there was some safety net underneath the market."[41]

Meanwhile the stoic Hans Tietmeyer, the president of the German Bundesbank, was sounding the alarm. He had wanted the large Wall Street firms and banks to at least receive a "haircut" for their profligate investments; instead, they were reaping healthy profits. He warned that "excessively large financial assistance" might "undermine the proper working of the markets. Consequently, market players may easily underestimate the risks, trusting in government guarantees or a bailout by the international community. This moral hazard problem has to be taken seriously." As he wrote in the Bundesbank's 1997 annual report, "Although rescue operations of this kind may afford relief in the short run, for the future they involve the risk of a recurrence of unwelcome behavior on the part of market players."

Joint Economic Committee Chairman Jim Saxton, a Republican from New Jersey, warned that these bailouts "not only reinforce existing risk-promoting incentives" but "create incentives for additional risky lending by international financial institutions."[42]

Robert Rubin recognized the problem. As he noted in his memoir, "We did worry about the 'moral hazard' problem that had gotten so much attention during the Mexican peso crisis." But in the end, he came to the conclusion that it was not really a big deal. Focusing on short-term issues of stability and growth was more important. With masterful understatement, he remarked, "A byproduct of programs designed to restore stability and growth may be that some creditors will be protected from the full consequences of their actions."[43]

The next crisis for Wall Street was Russia, which had opened up to Western investment after the collapse of the Soviet Union in 1989. There was widespread acknowledgement that Russia was a risky environment. Fortunes would be made but could also be lost. This should have kept away the more sober Wall Street investors, and in the old days it probably would have. But the new Clintonian investor class had been encouraged to throw caution to the winds by the promise of bailouts and the promiscuous aura of risk promoted by Clinton himself in the name of forging a global economy.

Meanwhile, on Wall Street there was talk that a new firm had opened up shop. They called it "Government Sachs." The Clinton administration was now joined at the hip with the large firms and seemed to want to ensure that their investments did not fail. So other investment houses and large banks began looking at where Goldman Sachs was investing and following suit, confident that they would have added "insurance" from risk.

As Andrew Ross Sorkin recently noted in the *New York Times*:

Back in 1998, the troubled Russian government was able to borrow $1.25 billion in the international capital markets when it was having great difficulties meeting its domestic obligations. I recall asking one money manager why he put up some money, given Russia's obvious problems. There was, he assured me, no real risk. The Russian bonds were underwritten by Goldman, which would not have gotten involved without getting an assurance from Robert Rubin, the former Goldman chief executive who was then President Clinton's Treasury Secretary, that the American government would step up if needed to prevent a Russian default." [44]

Russia, like Mexico, had become hooked on short-term foreign capital. The Kremlin needed foreigners to buy large quantities of GKOs, which were high-risk, high-yield bonds that offered returns in the 20 to 30 percent range and had a short maturity, usually only three months. The problem was that GKOs could be bought with dollars but the interest and principal were paid back in rubles. Few were certain whether the ruble would have much value in the future given the faltering Russian economy. As a result, the price of GKOs would soar or plunge depending on the state of investors' beliefs about whether the government would actually make the payments. In a matter of days they might plunge 30 percent.

Once again, big-time gamblers at foreign hedge funds, brokerage firms, and commercial banks jumped in and were holding $20 billion of these highly combustible bonds by the spring of 1998. The investors knew the risk they were taking. The strategy of investing in Russian debt securities was actually referred to as a "moral hazard play" by many investors, which suggests that the Clinton administration's repeated bailouts had indeed influenced them.[45] Many of them had used arbitrage, borrowing heavily to finance their purchases. This is a common practice with some investment houses because the overall return on the bonds almost always exceeds the interest on the loans. But the bonds are pledged as collateral, and when they plummet in value, as the GKOs soon did, creditors demand repayment.[46]

Big Wall Street investors had adopted the attitude that Russia was "too big, too nuclear to fail," so they were quite comfortable throwing good money into bad investments, believing that if necessary the Clinton administration would ride to their rescue again.

Clinton did not disappoint. An international bailout package of $22.6 billion was shortly unveiled, with U.S. taxpayers on the hook for almost a third. Goldman Sachs helped the Russian Finance Ministry with the underwriting of a new round of equally risky bonds, this time offering payment in U.S. dollars. Soon, however, the Clinton administration's plan was subverted by the corruption of Kremlin officials. About a month later, the Kremlin announced that it was devaluing the ruble and effectively defaulting on its Treasury bills. Much of the money from the bailout was assumed to have ended up in the hands of corrupt Russian leaders and their offshore bank accounts. As Charles Dallara of the Institute of International Finance in Washington explained, "you had the West, the I.M.F., Western banks and Western governments pouring money in the front door, and a select group of Russian citizens taking it out the back door." Still, some of the major players managed to get out of Russia unscathed thanks to the bailout.[47]

One of those who did not fare so well was a New York-based hedge fund called Long-Term Capital Management (LTCM). In September 1998, the Russian default triggered a series of events in the financial world that threatened the firm's survivability. Run by a Wall Street veteran, John Meriwether, with two Nobel Prize–winning economists as partners, it used sophisticated mathematical models and promised impossible returns year after year. Open to investors who could cough up $10 million per year, it was a highly leveraged venture. LTCM was making investment bets leveraged 30 to 1, meaning that it was borrowing $30 for every $1 it invested. If things went well, that kind of leverage could make it a lot of money. But if things went poorly, it could be in serious trouble very quickly.[48]

Much as they had done in Mexico, South Korea, Thailand, Indonesia, and Russia, big firms lined up to provide credit to LTCM with few questions asked. Citigroup, Goldman Sachs, Merrill Lynch, Chase, Bear Stearns, and Morgan Stanley put up money with little understanding of how the secretive fund even worked.

But with the Russian default, the markets were down and the highly leveraged LTCM was imploding. Robert Rubin and his cohorts at Treasury began sounding the alarm bells again, just as they had during every other financial crisis they had encountered over the course of the past five years. Collapse, they said, was not an option; it might lead to "contagion" and spread throughout the increasingly integrated global financial system. But there was not much evidence to prove this was the case. The most Rubin could offer was that LTCM's collapse might mean that "businesses and consumers" would find "credit less available and more expensive." He admitted that the collapse would not lead to "systemic disruptions."[49] Hardly the stuff of financial apocalypse.

But it was certainly affecting the big firms on Wall Street. Goldman Sachs was now delaying its initial public offering, and cochairman Jon Corzine would resign.[50] (Corzine had run up quite a string of bailouts, from Mexico to South Korea and now Wall Street itself, in a matter of just a few years.)

With the approval of the Clinton Treasury Department, New York Federal Reserve President William McDonough convened the heads of the big investment firms at the Fed's New York headquarters and the full-court press to secure an infusion of cash for LTCM began. The number of Clinton-era Wall Street bailouts was getting so large that those working on the LTCM case "came in for some ribbing at the Treasury's senior staff meeting from department colleagues who had been principally responsible for overseeing the rescues in Asia and Russia." Timothy Geithner, who was then undersecretary of the Treasury for international affairs and later was appointed secretary of the Treasury by President Barack Obama, joked with Gary Gensler, the assistant secretary of the Treasury for financial markets, "So, you had to do a bailout too, huh?"

"Well," he retorted, "you guys were getting too much attention."[51]

As Rubin pointed out in his memoir, the purpose of the LTCM infusion was to secure payment for Citigroup, Goldman, Merrill, Chase, Bear Stearns, and Morgan Stanley—not to save the firm. "This capital infusion gave the hedge fund breathing room to liquid its positions in a more orderly fashion."[52] For good measure, the Federal Reserve cut short-term interest rates to calm investors' fears.

By the time LTCM was liquidated in 2000, all the key investors had recovered their investments, and some had even made a profit.[53]

This latest Clinton-organized bailout caused raised eyebrows in Asia, which had lately been lectured by Rubin himself about the dangers of crony-style capitalism. "It showed us that the Americans could easily ignore their own principles," remarked Tadashi Nakamae, a prominent Japanese economist. The reason for the bailout was not the risk of a systemic crisis, explained Nakamae. "Of course, if LTCM fails, there would be a lot of negative chain reactions. But to me, that has nothing to do with systemic risk. It is simply a failure of the speculative chain." Indeed it was. The speculators had guessed wrong and were now getting help from Washington *again*. Some paragons of capitalism, such as Citigroup, had now been bailed out *four times* by taxpayers for their imprudent and risky behavior.[54]

Professors Burton Malkiel of Princeton and J. P. Mei of New York University were even more direct in the *Wall Street Journal*:

> If unsuccessful hedge funds are not allowed to fail, if brokerage firms believe they will somehow be protected from the effects of far too liberal margin requirements, if banks believe help will be forthcoming should loans go sour during unsettled market conditions, how will we discipline future decisions of investors and lenders? Will such intervention make our financial system even more fragile later? . . . Anything that weakens the effect of market discipline and that lessens the punishment the market afford speculators when they have made incorrect decisions is likely in the long run to lead to more instability.

The consequences of Clinton-style state capitalism were clear: Wall Street would take even more risks. As Gary Stern and Ron Feldman of the Federal Reserve point out in their book *Too Big to Fail*, these sorts of bailouts created a moral hazard, in that these banks had less incentive to monitor risks than they would ordinarily. As they put it, this is not to suggest that bankers were sitting in a back room scheming to defraud the government. But they are rational actors and respond to signals they receive from the markets. And in the 1990s they were essentially told that in terms of risk-reward calculations, the optimal decision was to risk and risk big because the government would bail them out.[55]

In effect, many of the Clinton bailouts were guarantees that served as giveaways to rich investors who had gambled on high-yield deposits in shaky financial institutions. The investment bankers behaved recklessly and were saved from the consequences of their actions by the Clinton administration. To make matters worse, Robert Rubin knew this to be the case. In his memoirs he assigned "a significant share of the blame [for these financial crises] to private investors and creditors" who "didn't understand the risks and supplied too much money." Needless to say, this would encourage more reckless behavior in the future.

Argentina had a deadbeat reputation in international finance, having defaulted on four large loans in its history, most recently in the 1980s. But Wall Street firms, confident that they would be bailed out if they got into too much trouble, rushed right in. J.P. Morgan and Merrill Lynch started selling billions of dollars worth of Argentine government bonds, telling investors that they were a steady and safe investment. "Every time we finished a meeting [with portfolio managers] the orders would come in," recalled Miguel Kiguel, who was then Argentina's undersecretary of finance. "People were desperate to buy Argentine." The same sort of behavior had gone on in Brazil, which was now having trouble making its payments. Citigroup had $9.8 billion at stake and Chase Manhattan $4.3 billion in outstanding debt there by 1998.[56] Fleet Boston and J.P. Morgan also had billions of dollars in the mix. The Clinton administration helped arrange a $22 billion bailout for which taxpayers were on the hook for $5 billion. (Again, because of the pesky U.S. Congress, the Clinton administration had to tap that special U.S. Treasury fund.) Citigroup and the others happily declared that it was a great time to do business in Brazil. Days later it issued a report explaining that Brazil and the rest of Latin America were "ripe for investment." The fact that the fruit was being plucked from taxpayers' wallets didn't seem to bother them.[57]

With so many investment crises breaking out year after year, at one point Clinton even proposed that a "contingent credit line" (CCL) be offered to "help countries ward off financial contagion." Critics derided the plan as a sort of "credit card" for countries with bad financial habits. The Germans thought the plan was outrageous, and Klaus Regling, Germany's G7 deputy, attacked it as encouraging further irresponsible behavior. Eventually the notion was dropped due to international opposition.

On coming to office in 2001, the Bush administration tried to curtail these practices. Paul O'Neill, Bush's first Treasury secretary, complained that governments would ride in on a white horse and throw "money at everybody, and the private sector people get to take their money out." He also said that American taxpayers should not be bailing out Wall Street firms. "As we in the finance ministries of the world talk glibly about billions of dollars of support for policies gone wrong, we need to remember that the money we are entrusted with came from plumbers and carpenters who sent 25% of their $50,000 annual income to us for wise use."

John Taylor, who was then the undersecretary of the Treasury for international affairs, held similar views. What was needed, he said, was tough love, and in the words of another observer the Bush administration believed that "financial markets, like wayward children, sometimes needed harsh discipline rather than rescue."[58]

Reversing course completely proved difficult, however. The Bush administration, reported Bluestein, "did take a harder line than the Clinton administration" by "refusing to follow the Clinton example of using U.S. taxpayer dollars to fatten up the rescue packages and letting the brokerage firms get out with their profits intact." Concerned that it might shock the global system, it believed slow reform was necessary. But the damage had already been done.[59]

Imagine for a moment that you have a high-testosterone friend, with few if any moral principles, who is eager to score with as many women as possible. Now someone is offering him complete protection—no pregnancies and no sexually transmitted diseases. What do you think he will do?

Clinton offered that kind of protection. And we now know what Wall Street did with it. When the International Monetary Fund looked at the 2008 financial meltdown, it clearly saw how Clinton's actions had created complacency about risk on Wall Street. As its 2008 *Global Financial Stability Report* put it, financial firms became "more complacent about their liquidity risk management systems" and would rely on federal government intervention to solve any "liquidity problems" that might arise.[60]

The man at the center of these decisions was Robert Rubin, a thoroughly Clintonian figure who emerged in the 1990s as one of the most universally admired statesmen in the world. As Robert Kuttner noted, "In reviewing published articles on Rubin going back two decades, I literally could not find a single feature piece that was, on balance, unflattering." Yet Rubin enabled Wall Street to take greater and greater risks, refusing to allow them to take their lumps for reckless behavior.

Rubin's influence does not end there. His protégés now populate the top ranks of the Obama administration and were deeply involved in the most recent round of federal bailouts. Timothy Geithner is now Obama's Treasury secretary; Lawrence Summers is Obama's director of the National Economic Council.

The Clinton administration helped create a bailout mentality on Wall Street that led many of the large firms to ignore risks and chase higher yields, believing that the federal government would bail them out if they got into trouble. As the Berkeley economics professor Barry Eichengreen argued, "In Mexico in 1995, Korea in 1997, and Russia in 1998, official funds were used to repurchase and retire short-term debt that private investors were unwilling to hold. Having benefited from high interest rates while their money was in place, creditors were effectively protected from capital losses when it came time to sell. The moral hazard thereby created an obvious incentive to engage in even less prudent lending, setting the stage for still larger crisis and still larger bailouts."[61]

As we shall see, it was government meddling in financial markets that caused the economic crisis that engulfed us in 2008, when the campaign to make credit and affordable housing a civil right collided with the Clintonian culture of risk on Wall Street. Both were fueled by an activist liberal government that thinks it must constantly intervene in social and economic mechanisms in order to tip the scales toward equity and justice. The consequences of such meddling have been, and will continue to be, dire.

MINORITY MELTDOWN

A Tale of Two Bubbles

Owe the bank $100, that's your problem. Owe the bank $100 million, that's the
bank's problem.

—J. P. GETTY

The economic crisis that besieged the world beginning in the fall of
2008 seemingly came out of nowhere. But in reality it was a potent
mix of two emerging economic forces—a collision of two bubbles
created by government interference in the marketplace.

Over the course of thirty years, housing activists and their allies
in Washington had pushed to compel banks to make loans based not
on whether people could pay them back but as a civil right. This was
the subprime housing bubble. Beginning in 1998 and running up to
2007, they had managed to force banks, through the CRA, to commit
$4.2 trillion in such loans. Trillions more were committed outside
the CRA by mortgage lenders pressured by the threat of federal lawsuits to loosen their lending standards or eager to sell the mortgages
to Fannie Mae and Freddie Mac. By 2005, almost 33 percent of new

mortgages were interest-only and 43 percent of new home buyers put no money down. That was what the activists and their accomplices in Washington had always wanted. Now they had succeeded beyond the wildest dreams of Saul Alinsky and his acolytes.

The Federal Reserve's loose monetary policy kept interest rates low and made borrowing money cheap. Not only could you borrow to buy a home and put nothing down, you could pay a very low interest rate. Why not go into debt? This had the effect of opening the housing market to wild speculation; houses were bought and sold quickly, driving up prices artificially and creating what many decried as a dangerous bubble.

Meanwhile, the Clinton administration had created a culture of government-funded risk in the financial markets and in collusion with Wall Street had midwifed a new form of state capitalism. The constant bailouts of failing Wall Street firms meant that the large banks and investment houses were shielded from the consequences of unruly speculation. David Kansas of the *Wall Street Journal* asked the head of one bank, "Aren't you concerned about taking on so much risk?" His answer: "More risk is simply more profit." [1]

As the British philosopher Herbert Spencer once put it, "The ultimate result of shielding men from the effects of folly is to fill the world with fools." By the end of the Clinton administration, Wall Street had more than its fair share.

By giving trillions of dollars in loans to people who could not pay them back and encouraging Wall Street to speculate wildly in mortgage-backed securities, Clinton had helped to inflate two massive bubbles. It was only a matter of time before they burst.

The mainstream media and numerous pundits have explained the crisis as a logical consequence of capitalist greed and avarice. Americans went nuts, flipping condos in south Florida, greedily buying up homes they could not afford, and looking for a quick buck. Beginning in 2005–2006, when property values stopped climbing, the house of cards that was the real estate market began to teeter. In 2007, the cards began to fall as many people found themselves "underwater" on their property—that is, they owed more than it was worth. The problem was made worse because Wall Street investment houses and banks had bought up a lot of securities based on these loans and were caught with their pants down when the foreclosures started.

There is some truth in this account. There are plenty of empty condos in places such as south Florida, the leftover debris of flippers and speculators. And Wall Street did dive into mortgage-backed securities, looking for a quick buck. But this account ignores the central facts of what happened.

Consider the foreclosure crisis, which sparked the broader economic collapse. Ground zero for foreclosures is not the wealthy enclaves of south Florida or California, nor the middle-class neighborhoods of Omaha or Dallas. The true epicenter of the crisis was the poor and minority areas that lay at the heart of the activists' agenda. As we saw in chapter 3, there was an explosion of loans to these areas beginning in the mid-1990s. By 2005, the push for relaxed lending standards had reached absurd heights.

Catrina Roberts, a single mother of four, was told in 1999 that the best thing she could do was get out of her apartment and buy a house of her own. Never mind that she was taking in only $880 a month after taxes as a home health aide. So she secured a fixed-rate mortgage and put down less than 3 percent on a house in Cleveland. But then the front porch broke, and there were other repairs. Eventually she started to fall behind. Six years later, she lost the house to foreclosure. "I know when you buy a house, eventually you have to put work into it," she explained. "But I didn't know it would lead me here, because if I did I would have never bought it. So, I am at a point right now that I don't want to ever buy a house, ever again." [2]

Roberts was not alone. The Slavic Village section of Cleveland, once an enclave of East European immigrants and now home largely to blacks and Latinos, saw the same story replayed hundreds of times. By early 2006, 5 percent of the homes in the neighborhood were vacant. Those who managed to stay saw crime rates rise, property values fall, and neighborhood blight take over sections of their community.

In California, Alberto Ramirez and his wife, Rosa, non-English-speaking strawberry pickers earning $14,000 a year, had lived for years in county-assisted farmworker housing with their three children. Eager to move, they secured a mortgage from Washington Mutual to buy a home for $720,000. Needless to say, Ramirez quickly fell behind on his payments and defaulted on his loan.[3]

Then there are people like the ACORN activist Veronica Peterson. Protesting in Washington, D.C., she declared that subprime loans were "weapons of mass destruction" and claimed that she had been a victim of predatory lending. Peterson had taken out a $436,000 mortgage from Washington Mutual to buy a home in Fox Grape Terrace in Columbia, Maryland, in 2006. She then took out a second, "piggyback" loan of $109,000. Her total payments per month were $3,386.17. Peterson was foreclosed on in July 2007. The foreclosure revealed that there was $435,735.86 due on the first mortgage. Unlike Catrina Roberts or Alberto Ramirez, who had at least tried to make their payments, Peterson had made just one payment over the course of a year—that was it. She had essentially lived rent free for a year. Still, the *Washington Post* and ACORN declared her to be a "victim" of the mortgage crisis.[4]

In sum, those hardest hit by the financial crisis are the very poor and minority neighborhoods that activists were claiming to help. Studies show that wherever you look—New York, California, Florida, Ohio, Michigan, or Minnesota—minorities are far more likely to lose their homes to foreclosure. And this sweeps up other individuals in the storm because it creates blight in their neighborhoods. "People try to sell houses in these pocket subdivisions and they can't sell because of foreclosures," said appraiser David Turner in

Ohio. "They're forced to drop their price and the neighborhood suffers."[5]

The simple truth is that people like Catrina Roberts and Alberto Ramirez would have been better off if the government had not encouraged such risky loans. Their credit would have been better, and they would have had the prospect of owning a home in the future, when they were better prepared.

In 2008, a massive study by the Boston Federal Reserve Bank—the same bank that had helped kick off the subprime lending craze with its flawed 1992 study—looked at hundreds of thousands of mortgages and foreclosures and discovered that "in the current housing crisis foreclosures are highly concentrated in minority neighborhoods." The study noted that this was a unique phenomenon, "even relative to past foreclosure booms," such as that of the early 1990s. Over a twelve-year period from 1990 to 2007, the study found, subprime borrowers in poor and minority neighborhoods were *seven times* more likely to lose their homes than the general population. In short, thanks to the efforts of the activists and their allies in Washington, the health of the American financial system came to rest on individuals who either didn't know how to handle debt or didn't care.[6]

Another study by the Pew Research Center found the same thing, namely, that the foreclosure crisis was concentrated in poor and minority areas. The study discovered that the most significant determinant of the rate of foreclosure was the "immigrant share of the population" and the "native-born Hispanic homeownership rate." In short, the very neighborhoods that had been targeted for more loans by CRA activists and Countrywide Financial—egged on by Fannie Mae and the Congressional Black and Hispanic Caucuses—had the highest foreclosure rates. In Miami-Dade County, Florida, and Sacramento County, California—two counties with some of the highest foreclosure rates in the country—foreign-born immigrant status was the largest factor in influencing foreclosure rates.[7]

An analysis by the *New York Times* echoed these findings. The paper discovered that in areas where default rates are at least double the regional average, 85 percent of the neighborhoods are majority black and Latino. Minority mortgage holders had a foreclosure rate some three and a half times higher than the national average. When the paper analyzed trends in New Jersey, it found that foreclosures "hit hardest" in the poor urban areas of Newark, Paterson, Elizabeth, and Willingboro. And they discovered that black census tracts had four times the foreclosure rate of white tracts. In Connecticut, the paper found the same trends: poor and black census tracts had foreclosure rates that were three times greater than others.[8]

In Los Angeles, Reuters found that zip codes with high poor black populations had high foreclosure rates, "a common combination," the news service reported. In Ohio, the Kirwan Institute for the Study of Race and Ethnicity at Ohio State University found that "communities with high concentrations of minorities were hit hardest by the foreclosure crisis."[9]

Though no clear numbers have been published that show the current default rates for CRA loans, the Bank of America did reveal in 2008 that while CRA loans represented 7 percent of its loans, they accounted for a staggering 29 percent of its losses.[10]

Some housing activists and members of the Obama administration want to claim that these high foreclosure rates in poor and minority neighborhoods were brought about because borrowers were saddled with adjustable-rate mortgages or subprime loans that they did not need. For example, Obama's secretary of Housing and Urban Development, Shaun Donovan, claims that roughly 33 percent of the subprime mortgages given out in New York City went to borrowers with credit scores that should have qualified them for conventional loans.

But this is a myth that is designed to shift blame to the banks rather than to the activists—and their government enablers—who pushed the poor to take out loans they were clearly not prepared for. As the Boston Federal Reserve Bank found in its study of the crisis, the idea that individuals were coaxed into subprime loans they did not need is a myth. Credit scores alone don't determine whether you get a subprime loan. The Boston Fed found that many of those with good credit scores failed to provide specific information about income, assets, or other important information that would determine what sort of loan they would get. Subprime loans did not explain the high rate of foreclosures in poor and minority neighborhoods: blacks suffered foreclosure rates three times those of whites and Hispanics twice the rate of whites, even when they had the same kinds of loans. Blacks on average had put less money down, had lower incomes, and had taken on more debt. But they were given loans because of the flexible underwriting rules activists and their allies in Washington had been pushing for three decades.

The Boston Fed study also exploded the myth that the problem was adjustable-rate mortgages. This argument holds that people lost their homes because they unwittingly took out adjustable-rate mortgages and when their payments went up they lost their homes. But the Boston Fed found that in a majority of cases, people who had adjustable-rate mortgages and were foreclosed on lost their homes *before* the rate was adjusted.[11]

There is simply no evidence that minorities and the poor were targeted by subprime loans. The New York Federal Reserve Bank looked at more than 75,000 adjustable-rate mortgages and found that minorities did not pay higher rates. Indeed, the study concluded that "minority borrowers appear to pay slightly lower rates, as do those borrowers in zip codes with a larger percentage of black or Hispanic residents."[12]

The problem lies not with the lenders who nefariously pushed subprime mortgages but with the flexible underwriting standards that had been foisted on them by activists and politicians in Washington. As Professor Stan Liebowitz of the University of Texas reported, 51 percent of all foreclosed homes had prime mortgages. By looking at loan-level data covering more than 30 million mortgages, Liebowitz found that "by far the most important factor" in determining whether buyers went into foreclosure was whether they had put any money down on their home or had equity in it. In other words, the culprit is the relaxed lending standards championed by the Left for some thirty years: no-money-down loans and liberal underwriting.[13]

Despite all the talk about helping poor and minority communities prosper through this grand liberal social experiment, many of the very communities that were supposed to benefit are now devastated. As one writer put it, "For black America, the 'mortgage' meltdown looks less like a market hiccup than a massive strip-mining of hard won wealth, a devastating loss that will betray the promise of class mobility." Those who played by the rules now find themselves in neighborhoods filled with shuttered homes and broken windows. In short, the activists' dream has become a nightmare for those they professed to be helping. "Rather than helping narrow the wealth and ownership gap between black and white," admitted Colvin Grannum of the Bedford Stuyvesant Restoration Corporation, "we've managed in the last few years to strip a lot of equity out of black neighborhoods."[14]

It is hard to avoid the conclusion that the people in these neighborhoods would have been much better off if the government had not tried to help them. The mortgage crisis—and especially the meltdown of minority neighborhoods—is directly related to well-meaning efforts by liberals in government to tilt the housing market in their favor. The toxic derivatives bubble that arose was a rational response to this governmental tampering, while the failure of the market to discipline those risky instruments was likewise a result of the insular investment climate fostered by Clinton and his economic team.

Yet this is one lesson the Clintonian liberal technocrats never seem to learn from the catastrophic messes they create. The absolute limits of government's power to "do good," to "spread the wealth" and "lift people up," to realize someone's vision of a just society, never occur to them. Instead it is always just a question of "smarter" management.

Some financial analysts saw the debacle coming. In October 2005, Meredith Whitney, a well-respected financial services research analyst formerly with Oppenheimer & Co., declared that the prospects of an economic earthquake were real. Moreover, she saw the culprit as the push for greater home ownership. Whitney noted that up until 1994, home ownership levels in the United States had remained constant at about 64 percent. "That level changed fairly dramatically after 1994, when the standards by which an individual could qualify for a mortgage and the required down payment levels became much more liberal. Due to such changes in guidelines, 1.5 million new homeowners were created and homeownership rates rose relatively steeply to 69% where they stand today.... The incremental 5% of new homeowners would not have qualified for a mortgage prior to 1994." Those new home owners got their houses due to subprime loans. As she wrote, "Since 1996, sub-prime lending has grown 489%—from $90 billion to $530 billion—largely through the extension of credit to first-time borrowers." [15]

Beginning in 2006, those loans began defaulting in large numbers. At lenders such as Countrywide it was not long before almost 20 percent of mortgages were seriously late in being paid. Those defaults led to foreclosures, which in turn threatened the survival of highly leveraged lenders such as Countrywide, Washington Mutual, and Ameriquest. The wave of defaults soon rippled throughout the financial world because many of the mortgages had been sold to investors in the form of mortgage-backed securities. Between 2002 and 2007, nearly $2 trillion in mortgage-backed securities had been issued in the United States. A lot of them went through Fannie and

Freddie and were assumed to be backed by the federal government. Pension funds, hedge funds, and investment houses all held large sums in mortgage-backed securities. They now began to feel financial pressure as the people at the bottom of the ladder stopped making their payments.

Mortgage-backed securities are very complex investment tools. Basically, they involve pooling large batches of mortgages, splitting them up into small pieces, and selling the resulting products to investors as securities. As financial journalist Dave Kansas points out, Wall Street had embraced mortgage-backed securities for years because they were based on real estate, a tangible asset that had innate value. They also typically had a long payback period; very few people paid their mortgages off early. But such instruments are only as safe as the underwriting standards that go into them. When they were first created back in the 1970s, mortgage-backed securities had very strict underwriting standards. Old-style mortgages with 10 percent down and strict requirements for income, assets, and credit history made them relatively safe investments. But those standards changed, largely because of prodding by Washington. As we have seen, the Clinton administration proudly took credit in 1998 for encouraging the explosion in subprime mortgage-backed securities and pushing banks to "revis[e] their underwriting practices, making lending standards more flexible." One has to wonder: are the people involved still so proud?[16]

The big banks and Wall Street firms were also highly leveraged, investing with borrowed cash. Investment banks might borrow $25, $30, or even $35 for each dollar they actually held. They were also leveraged because they bought derivatives. They had done the same thing during the Asian crisis in the 1990s and had walked away with their profits intact; why not assume they could do so again?

"The range of derivative contracts is limited only by the imagination of man, or sometimes so it seems, madmen," Warren Buffett once said. In 1993, the so-called notional value of derivatives was $14 trillion. By 2001, it had blossomed to $100 trillion. As Robert Rubin recalled during the Asian credit crisis, "The commercial and investment banks had no precise idea of exposure to South Korea in financial derivatives. It took them a week to find out." But the Clinton administration made them whole with a bailout, so Wall Street didn't even need to stop and consider the dangers of what it was doing. Now the large investment firms were in the same boat again with derivatives. They didn't know how much they were exposed. In the end it was discovered that among them, JPMorgan Chase, Lehman Brothers, and Bear Stearns had derivatives exposure of some $100 trillion.

What's more, in many instances the banks and investment houses now asking for a bailout had been bailed out before for the very same reason. This was not something that afflicted the entire financial system. As Bloomberg News noted, "Ninety percent of the trades [in derivatives] were concentrated in the hands of 17 banks, according to the Federal Reserve Bank of New York." Which were these banks? As you may have guessed, most were the same outfits that had been bailed out several times already.[17]

Banks and mortgage lenders also became more aggressive in the sorts of loans they were writing because Fannie Mae and Freddie Mac needed to buy risky mortgages to meet their goals of providing trillions of dollars in loans to poor and minority applicants. This has been portrayed by the Left and its media allies as evidence of capitalist greed and exploitation run amok. In reality it was a rational response to government-caused distortions in the market. As the financial writer Dave Kansas notes, "With Fannie and Freddie aggressively acquiring risky mortgages, regular banks had to get more aggressive themselves. This led to a beggar-thy-neighbor environment that helped undercut mortgage-discipline throughout the country."[18]

As the real estate boom came to an end in 2006 and the holders of mortgage-backed securities started feeling the pinch, hedge funds began to fail. In 2007, a fund run by the French bank BNP Paribas went under and another run by Bear Stearns failed. Both had huge holdings in real estate securities, many of them in derivatives.

By early 2008, the subprime mortgage lender Countrywide was in free fall and Angelo Mozilo sold the company to Bank of America at bottom-of-the-barrel pricing. In March, Bear Stearns, suffering huge new losses on its real estate investments, was sold to JPMorgan Chase for pennies on the dollar. The $29 billion rescue deal was orchestrated by the federal government and was designed not only to save Bear Stearns but also to protect JPMorgan Chase, which held a large number of derivatives linked to the company.

By July, IndyMac Federal Bank went under. Cofounded by Angelo Mozilo, the bank had been spun off from Countrywide in 1997. There was a minirun on the bank after Senator Charles Schumer released a letter he had sent to the FDIC outlining his concerns about the solvency of the bank. Like Countrywide itself, IndyMac was swamped by mortgage losses.

One week after IndyMac failed, Treasury Secretary Henry Paulson had to step in and orchestrate a government takeover of Fannie Mae and Freddie Mac. Based on the goals established by Jim Johnson and Franklin Raines, the GSEs had a couple of trillion dollars tied up in high-risk mortgages to poor and minority applicants with a history of failing to make loan payments. At that point Fannie and Freddie were packaging 75 percent of all mortgages in the United States. The federal guarantee had always been implicit; now Paulson made it explicit. The American taxpayer was on the hook for more than $5 trillion in mortgages. By the end of 2008 Fannie was reporting quarterly losses of $29 billion.

The failure of Fannie and Freddie was a watershed. Two weeks after the takeover the crisis deepened when so-called credit spreads exploded. Credit spreads are options based on the difference between interest rates paid by U.S. Treasury bonds and other kinds of debt. The wider the spread, the higher the fear in the market. As Meredith Whitney observed, "A company is only as solvent as the perception of its solvency." By mid-September the spread broke all previous records.[19]

Suddenly it became very expensive for businesses to borrow from banks, largely because of the fear and uncertainty in the financial market. Large companies, which had solid balance sheets, found it hard to sell bonds. What had begun as a real estate crisis and morphed into a financial crisis was now a general crisis affecting the entire economy.

A week after Fannie Mae and Freddie Mac went into conservatorship, Lehman Brothers filed the largest bankruptcy in American history. The venerable investment house soon collapsed under the weight of bad investments in mortgage-backed securities and derivatives. Lehman's demise caused enormous problems in the banking system because it impacted hundreds of billions of dollars in credit default swaps that were being underwritten by the insurance giant American International Group (AIG). The next day, the federal government had to step in and offer $68 billion to help AIG stay afloat, in exchange for 80 percent of the company. AIG had huge holdings in credit default swaps, and with investment houses suffering huge losses and some going under, it was on the hook for tens of billions of dollars it simply did not have. AIG also had more than $30 billion worth of subprime mortgages on its books in the form of housing derivatives and an equal amount of Alt-A securities, barely above subprime.

Soon Washington Mutual, the bank that had aggressively pushed into what it called multicultural lending, began to fail and was bought by JPMorgan Chase. Then Citigroup began to reel from its hundreds of billions of dollars in toxic assets, mainly in the form of subprime mortage-related derivatives. The collapse required an injection of $20 billion of taxpayers' money (eventually rising to $45 billion). Ultimately the federal government would guarantee $306 billion in troubled home loans and subprime bonds that the bank had written. It was a tremendous deal for Citigroup. In exchange, the federal government received $27 billion in Citigroup preferred stock, which pays 8 percent interest. In November 2008 the bank announced 52,000 layoffs. It was the fifth taxpayer-funded bailout Citigroup had enjoyed in fifteen years.

In a very real sense, the big investment houses had been walking a tightrope with a federal safety net under it. Back in 1991, Senator Christopher Dodd of Connecticut, the chairman of the Subcommittee on Securities, Insurance, and Investment of the Senate Banking Committee and a reliable friend of the Wall Street establishment, had inserted a few words into a bill whose primary purpose was to reform the Federal Deposit Insurance Corporation. The sentence, requested by Goldman Sachs and several other Wall Street firms, gave the Federal Reserve emergency powers to rescue financial firms that might be at serious risk. (This seems a subtle sleight of hand whereby the government, in assuming the power to rescue financial firms, imperceptibly assumed the responsibility for doing so.) This was the authority that the Fed would use more than a dozen times in 2008–2009, when Bear Stearns, AIG, Goldman Sachs, and Morgan Stanley faced imminent collapse.[20]

When the financial crisis erupted in full form in the fall of 2008, national panic erupted. Information was hard to come by. Banks were failing, and the stock market was in free fall. The Bush administration rushed through Congress the Troubled Asset Relief Program (TARP), which pumped some $700 billion into the economy in the hope of stabilizing the situation. Fed Chairman Ben Bernanke told a congressional committee that if taxpayer money were not made available to large financial institutions, "we may not have an economy on Monday."[21]

There is now considerable evidence that the crisis might not have been as widespread as many thought. An analysis by economists at the Minneapolis Federal Reserve Bank looked at bank and interbank lending and found that in fact there was no systemic crisis, as many in Washington had claimed. The vast majority of banks were lending money; the problem was concentrated among a few large players—the same ones who always seemed to be getting into trouble. Celent Communications, a Boston-based financial consultancy, came to the same conclusion and said that any liquidity problems were concentrated among a few large actors. Their names, of course, should by now be familiar. They include the large firms bailed out repeatedly over the past fifteen years.[22]

TARP was a $700 billion bailout package ostensibly designed to remove toxic debt from the balance sheets of large banks by buying it up. The feds pledged $600 billion to purchase mortgage-backed securities guaranteed by Fannie Mae and Freddie Mac. By the end of 2008, the federal government had pledged more than $7 trillion in loans and guarantees to shore up the financial system.

The economic tsunami that engulfed the United States—indeed, the whole world—by the end of 2008 was a culmination of these two powerful forces, both of which had been created by governmental tampering in the mechanisms of the free market in order to advance liberal ends. In pushing this agenda, liberal activists inside and outside government saw themselves as doing something virtuous: extending the gains of the civil rights movement to new groups of people farther down the economic ladder. To that extent, their motives were laudable and idealistic, at least in their own eyes. But the results have been disastrous, especially for the minority neighborhoods they meant to help. Much as baby boomers tried to improve American culture and left it in shreds, they have now done the same to America's economic system.

The great irony is that those who unleashed this economic calamity appear to be the main beneficiaries of the crisis they helped to create. What's more, they are back in power today and are busily planning the next one. Therefore we now turn to the Obama administration—the most ambitious, most activist left-wing government ever to take power in the United States.

ROBIN HOOD COMES TO THE WHITE HOUSE

How Barack Obama's Plans to Use Liberal State Capitalism Will Create the Next Big Economic Meltdown

It has often been said that the definition of insanity is doing the same thing over and over, expecting a different result. Government intervention, bailout capitalism, and liberal social engineering got us into this economic mess. The prescription now being offered to get us out of it? More of the same.

Without government intervention in the economy, the economic collapse would not have occurred. We would not have seen the explosion in subprime lending and the dramatic reduction of lending standards. Fannie Mae would not have been allowed to inflate the housing bubble, and subprime lenders such as Countrywide would not have grown into the toxic behemoths they became. Without government intervention in the form of the Community Reinvestment Act, banks would not have been compelled to extend trillions of dollars in loans to people with horrible credit. Without government intervention the large investment houses would have learned fifteen years ago that taking foolish risks, such as investments in toxic mortgage-backed derivatives, would cost them dearly in the end.

This grand social experiment has caused millions of Americans to suffer. Yet there is no evidence that lessons have been learned by the powers that be in Washington. To the contrary, the very people who steered us on the current course are back in power, tasked with cleaning up the mess they themselves made. Even more worrisome, after two previous attempts to insert the federal government into commercial markets, both of which ended in failure, liberals in the Obama administration are getting ready to do it again, on multiple fronts and on a grander scale than ever before.

In a word, the Robin Hood gang has decamped from Sherwood Forest and occupied Nottingham Castle.

Let's start at the top: Barack Obama, the activist and lawyer who fully supported the radical housing agenda, now sits in the Oval Office. Far from honestly looking at what got us into this mess, he blames everyone but himself—the banks, the lack of regulations, and greedy Wall Street executives. His plans to socialize health care in America represents a massive governmental intrusion in the market and will have tremendous distorting effects, creating artificial scarcity and driving up costs, which Obama expects to defray by imposing a massive surcharge on the wealthy. This in turn is entirely consistent with candidate Obama's definition of economic justice: take from the rich, give to the poor.

Meanwhile, the Obama administration is filled with old Clinton hands who were deeply involved in creating both the subprime mortgage and toxic derivatives bubbles. The new White House chief of staff is Rahm Emanuel, a prime enabler of this disaster when he served on the board of Freddie Mac. As a key Clinton staffer and later as a member of Congress, he was also a champion of aggressively enforcing the CRA and of Wall Street bailouts. (Emanuel had deep ties to Robert Rubin and Goldman Sachs, where he had served as a liaison between the financial giant and the Clinton fund-raising operation before entering the Clinton White House and launching his own political career.) Larry Summers, a Rubin protégé who succeeded him as Clinton's Treasury secretary, is now the president's chief economic adviser. Another Rubin protégé, Timothy Geithner, is Treasury secretary. All played key roles in the series of massive government bailouts and the enormous stimulus bill that were passed during Obama's first hundred days. What can be expected from this cast of liberal technocrats who revolve between Wall Street and Washington but more of the same?

Meanwhile, left-wing housing activists have found a niche at lower levels in the Obama administration.

At a December 2007 forum called the Iowa Heartland Presidential Forum, organized by former ACORN legislative director Deepak Bhargava, then-Senator Obama told a group of activists that he would be "calling all of you in to shape the agenda" were he to be elected president. Bhargava, who as an ACORN lobbyist pushed for the CRA and helped lay waste to the economy, was clearly pleased.

President Robin Hood has not disappointed. Roberta Achtenberg, the activist and Clinton assistant HUD secretary who went after banks in the name of fighting racism, was on President Obama's transition team for the Department of Housing and Urban Development. The individual who took Achtenberg's old job as assistant secretary of fair housing and equal opportunity is another activist, John Trasviña, the president of and lawyer for the Mexican American Legal Defense and Education Fund (MALDEF). At MALDEF, Trasviña has been a strong proponent of the exact sort of policies that got us into this mess.

Over on Capitol Hill, Congressman Barney Frank, the chairman of the House Financial Services Committee, pushed the affordable housing agenda, vigorously defended the CRA, staunchly protected Fannie Mae and Freddie Mac from reform efforts by the Bush administration, and supported the Clinton bailouts of Wall Street investment houses in the 1990s. Yet he told an audience at Boston College that he could have prevented the crisis if he had been given the opportunity. His brilliant solution? Give federal officials such as himself even more power to micromanage the real estate market. Frank recently pushed a bill that would allow the federal government to provide $300 billion in new guarantees to refinance mortgages and wanted another $100 billion in grants to communities to buy foreclosed homes. Oh, and by the way, he wants Fannie and Freddie to relax their lending standards even more. Sound familiar? [1]

Frank also supported efforts in March 2009 by Congresswoman Eddie Bernice Johnson, Democrat of Texas, to *expand* the Community Reinvestment Act to force credit unions and mortgage companies to comply with the law. [2]

Frank does not appear to be particularly worried about the cost of all this. "I think there are a lot of very rich people out there whom we can tax at a point down the road and recover some of this money." [3] This is the Robin Hood mentality at work.

Meanwhile, congressional Democrats who siphoned off campaign contributions from Fannie and Freddie, supported the affordable housing goals of the federal government, endorsed the CRA, and were all for bailouts of Wall Street investment firms have passed legislation to investigate what caused the economic crisis. As if they didn't know! This so-called Financial Crisis Commission will have subpoena power to "expose the roots of the current crisis." Senator Harry Reid, a Democrat from Nevada, and House Speaker Nancy Pelosi, a Democrat from California, will each name three commissioners. Any predictions as to what those hearings will find?

Clinton's HUD director Andrew Cuomo, who pushed the federal government more deeply into backing subprime loans and pushing relaxed lending standards to unprecedented heights (or depths), is now the grandstanding attorney general of New York. No doubt preparatory to a renewal of his moribund political career, Cuomo posed as an avenging populist, forcing Wall Street fat cats to return their year-end bonuses and promising to ensure that subprime lenders face justice. Never mind that he helped make it all possible during his tenure at Fannie Mae.

Not to be outdone in brazen hypocrisy, the very activist groups that once sued banks for failing to make enough loans in poor and minority neighborhoods are now suing those same banks because they loaned too much! Thus the NAACP has filed a class-action lawsuit accusing the banks of "reverse red-lining."[4] ACORN has done the same and demanded that Congress go after lenders and "hold them accountable." Never mind that ACORN has been forcing banks to move in this direction for close to thirty years.

In a similar vein, the National Community Reinvestment Coalition, the granddaddy of the CRA groups, argues that the problems of racism in lending persist, complaining that they are being targeted with low-cost mortgage products they cannot afford.[5] By a magical sleight of hand, yesterday's financial segregationist has become today's predatory lender. They must be wondering what hit them.

Needless to say, the enablers of these groups in Congress have parroted the same twisted narrative. As our old friend Congresswoman Maxine Waters explained on the House floor, "We need to prevent the now widespread practice of getting people into loans they simply can't afford." What's the moral of this story? Whether you get a loan or not, it's always the banker's fault.[6]

Jesse Jackson—as adroit a political trimmer as ever there was—has been pushing the same line. Although he strongly supported no-money-down mortgages in the 1990s, he now declares that the problem all along has been greedy lenders trying to hoodwink an unsuspecting public. In more sober moments, Jackson has admitted that the problems of poverty and home ownership in America are about something besides racism. "Many poor people are poor because their habits are poor," he has said. "The rich have rich habits. To save and invest is a rich habit. To spend and waste is a poor habit. We have to learn early on to have 'squirrel sense.' Squirrels are not known to have brains. There are no PhD squirrels, no attorney squirrels . . . but there are no homeless squirrels! Squirrels, by whatever instinct, put away some acorns for winter."[7] This was good advice when he offered it in the *Wall Street Journal* in 2001. But now he is clinging to the ugly canard that the subprime crisis is all about racism and greedy banks.

As the crisis began to unfold, activist groups held a meeting in Washington and declared that the Community Reinvestment Act "is still sexy after all these years." With a Latino community leader from Wichita, Kansas, dressed up as the Statue of Liberty, they discussed how to expand the reach of the CRA. Although the mandates imposed on banks through the CRA are very much alive, activists now want to have them cover other lenders.[8]

Despite the devastation wrought by this misguided activist crusade, the Obama administration is not backing down. If the Federal Reserve was initially sympathetic to banks during the crisis as they struggled to meet CRA requirements, the administration announced in March 2009 that that would no longer be the case. "We don't view the current credit environment as a free pass in terms of a bank meeting its CRA requirements," Timothy Burniston, an assistant director of the Federal Reserve, told the Consumer Bankers Association in March.

Just how twisted has the CRA become? East Bridgewater Savings Bank in Boston was actually slapped with a "needs to improve" rating by the FDIC during the midst of the financial crisis. The small bank had a cautious approach to lending, which had helped it weather the financial storm. "We're paranoid about credit quality," said bank president Joseph Petrucelli. East Bridgewater had no foreclosures, no bad or delinquent loans. But the FDIC evaluated the bank under the CRA and determined that it was not making enough risky loans. Just what we need in America—more risky loans.[9]

Activist groups have been warmly supportive of the Obama administration's policy initiatives. In February 2009, when President Obama announced his plan to pour tens of billions of taxpayer dollars into a plan to deal with foreclosures, ACORN was delighted. CEO Bertha Lewis said that Obama's plan "will finally put the full power of the federal government behind homeowners trying to pay the mortgage and neighborhoods reeling from the crisis."[10]

Congressional enablers tried to use the bailout as an opportunity to lavishly fund the very activist groups that had pushed these toxic policies in the first place. During the debate over the stimulus bill, Democrats pushed legislation that would have provided $4.19 billion for "neighborhood stabilization activities." Democrats wanted the funds to be available not just for state and local governments but for nonprofit organizations such as ACORN or Gale Cincotta's National Training and Information Center (NTIC). These organizations are already federal grant recipients and would likely be available for a large chunk of these funds.[11] Ironically, the NTIC was in the midst of settling a federal investigation over the misuse of a previous federal grant. Apparently the organization had used $207,000 in federal grant money . . . to lobby the federal government for more grant money. The grant from the Department of Justice was supposed to be used for Community Justice Empowerment programs. It settled with the U.S. attorney of the Northern District of Illinois for $550,000 in June 2009.[12]

The activists continue to use their Alinsky-inspired tactics to push their agenda. And the Obama administration has clearly been influenced by them. Whereas FDR was calm and reassuring during the depths of the Great Depression, Obama described the economy as on the brink of catastrophe and used the ensuing panic to push for a radical transformation of the economy. He has also used Alinsky's method of demonizing those who stand in the way of his agenda. When Chrysler filed for bankruptcy, the White House came up with a plan that would require creditors who bought corporate bonds to forgive 67 percent of the debt owed to them ($4.6 billion) in exchange for a 10 percent stake in Chrysler. The United Auto Workers (UAW), on the other hand, would have to give up 48 percent of the money owed to its retirees ($4.2 billion) but get 55 percent of the company. This was an abrogation of contract law; the bondholders were secured debtors, which meant they had to be paid before unsecured debtors (such as the federal government) that had poured money into the company or the labor unions. Yet on April 29, President Obama went on national television and singled out those creditors as bad actors who greedily wanted to profit from a bad situation. The creditors had actually said they would take 50 cents on the dollar for their investments.

The creditors went to court to protect their rights. And the White House responded with tactics that would have made Alinsky proud. Thomas Lauria, an attorney representing several of the creditors (which happened to be teacher pension funds in Indiana, among other bondholders) complained that the White House had called the bondholders "vultures" for insisting on their rights. Matthew Feldman, an attorney on the president's Auto Industry Task Force, called Lauria a "terrorist." The billionaire Steven Rattner, the head of the task force, apparently threatened another creditor that "the full force of the White House Press Corps would destroy its reputation if it continued to fight." The client withdrew from the case. The White House denies the charge.[13]

When executives at the insurance giant AIG received bonuses after receiving taxpayer money from the bailout, President Barack Obama compared them to suicide bombers. "It's almost like they've got—they've got a bomb strapped to them and they're got their hand on the trigger," he said. "You don't want them to blow up."[14]

Alinskyite techniques were also on full display when the White House summarily fired Rick Wagoner, the CEO of General Motors. The move prompted even the lefty Michael Moore to wonder, "Can he [Obama] do that?" There was no legal basis for it, but the president did it anyway. The day after, Treasury Secretary Timothy Geithner explained that the administration was open to firing other executives as well.

After the credit liquidity crisis passed, banks lined up to offer to pay back their TARP money with interest. But the Obama administration balked, saying it didn't want the money back. According to the Fox Business Channel's Stuart Varney, this was about control more than anything else. "By managing the money, government can steer the whole economy even more firmly down the left fork in the road," he wrote in the *Wall Street Journal*.[15] Eventually the Obama administration settled on a strategy of "stress tests" for banks. If the banks passed a government test, then they would be permitted to return the money. But it is the government that decides when banks can pay the money back—not the banks themselves.

Secretary Geithner and Larry Summers, the director of Obama's National Economic Council, are now in charge of economic policy. Both men, who were so instrumental in the Clinton Wall Street bailouts of the 1990s, are now pushing for more of the same. "No one should assume that the government will step in to bail them out if their firm fails," Geithner warned, even as the administration created that very impression. When Geithner unveiled his eighty-eight-page plan for financial reform in June 2009, he never even mentioned the issue. And the Obama team made the same mistakes concerning moral hazard as Clinton had. When the Obama administration poured $175 billion of taxpayers' money into AIG, the money actually went to a series of investors to whom AIG owed money because of bets on credit default swaps. Goldman Sachs received $12.9 billion through AIG and $10 billion in TARP money. What's more, Goldman received the money from AIG along with all of its profits. When Geithner was asked by Congressman Spencer Bachus, a Republican from Alabama, why taxpayer money was being used to pay off speculators who guessed right on credit default swaps, Geithner replied that the government did not have the authority to ask Goldman Sachs and other investors to take a haircut on their profits. (Of course, that was the very thing they had told Chrysler's bondholders to do.)

Just as in the 1990s, the investment houses took big risks and, courtesy of a government bailout, emerged with their profits intact. Little surprise that in the first half of 2009, Goldman Sachs gave its staff the biggest bonus payouts in the firm's 140-year history.

Successful capitalists are certainly entitled to their bonuses. But what do you call a system whereby you get to keep the profits but the government will help cover your losses? What kind of capitalism is that?[16]

President Obama has now also put taxpayers on the hook for loans to small businesses and individuals. Early in his administration he rolled out the Term Asset-Backed Securities Loan Facility (TALF) to encourage loans to small businesses. The idea is to loosen up credit by encouraging banks to pool together loans made to small businesses and households and then sell them to investors. But taxpayers would be guaranteeing those loans. As explained by Professor Elizabeth Warren of Harvard, chairman of the Congressional Oversight Panel, "Investors are putting up a small amount of the purchase price [for these loans] with the taxpayers putting up the money for the rest. If the securities go up in value the investors put a large amount of the profits in their pockets. If the securities decline in value investors can give up their part of the purchase price and leave the taxpayers stuck with the rest of the losses. Sounds like a great deal for investors."

Indeed it does—if not so great for taxpayers. But who cares about them? They exist just to finance government-sponsored investment schemes and pay the bills when they go bad.

With this type of state capitalism, or "financial mercantilism" as Kevin Phillips calls it, banks have seemingly few downsides for risky behavior. This sets the stage for future financial meltdowns even larger than the one we have just experienced.

For example, when Secretary Geithner saw to it that Citigroup and Bank of America each received $45 billion in TARP funds, the ostensible purpose was to "unfreeze" the credit market and get those banks lending again. The goal of the so-called Public-Private Investment Program (PPIP) was to get toxic assets off the balance sheets of the banks. But what did the banks do with those funds? They purchased more of the same toxic assets that had gotten them into trouble in the first place. They did so because they thought that buying them for 30 cents on the dollar was a good deal. And it *was* a good deal to take those risks—because they were using taxpayers' money to do it.[17]

This completes the political seduction of Wall Street, rescued again from its repeated risky behavior and confident that taxpayers will be there to bail it out again and again. *The Economist* magazine reported that by June 2009 the appetite for risk had returned. The "search for yield" is back.[18]

Nor have Wall Street executives shown much interest in taking responsibility for their risky behavior. Robert Rubin, who left the Clinton administration to assume a position as consigliere for the banking giant Citigroup, was the key force behind that bank's strategy to take on more risk. But when Citi faced financial collapse in the midst of the crisis, Rubin absolved himself of any responsibility. (Though if he had none, one can only wonder why he was paid $115 million by the company.) Just as he had done during the Clinton administration, Rubin tried to float above the crisis, declaring that the financial system as a whole was at fault.

Robert Parsons, a longtime member of the board at Citigroup, took over as chairman of the troubled bank and declared that the crisis was more or less everyone's fault. "Everybody participated in pumping up this balloon," he said. "Now the balloon has deflated." [19] Citigroup, which by June 2009 was substantially owned by the federal government, has now been bailed out *half a dozen times* by taxpayers over the past twenty years. This is not real capitalism.

There is a clash between the free-market interests of the American people and the short-term business interests of Wall Street. The liberal technocrats in the Obama administration have apparently assumed the role of mercantilists, favoring one sector over another, one firm over another, rather than simply encouraging an open and free market system. We are being told that economic mandarins will deliver us from our economic shortcomings, cushion us from calamities, minimize economic downturns. But their answer is always the same—taxpayers will bail everyone out. As Professor Thomas Cooley of New York University warns, "Every day we seem to move deeper into a world in which all the risk in the economy is socialized—it is being borne by taxpayers. By moving in this direction we are sowing the seeds of a much bigger crisis ahead." [20]

At the heart of the problem lies the American liberal Left and its continued inability to understand the nature of capitalism and the process of wealth creation. Saul Alinsky and the activists who pushed through the CRA recognized that by tapping the assets of private corporations they could accomplish great things that could not be achieved through the federal government. There simply wasn't enough money in federal coffers to do that. Today's activists continue to embrace this strategy even more than in the past.

As Professors Aaron Chatterji and Barak Richman of Duke University recently put it, "As progressives confront the problems of the twenty first century, be they global poverty and increasing income inequality, the scourge of HIV and other diseases, educational disparities, or climate change, an increasingly popular strategy is to enlist corporations in the effort." The strategy makes sense because "modern corporations have the financial resources, human capital, and global influence to advance progressive causes."[21]

This benign technocratic description masks a far more sinister reality. As Jonah Goldberg points out, fascism is not when corporations run the government but when the government enlists corporations in the realization of approved social ends.[22] One important means of doing this has been to create a concept of corporate citizenship that gives activists and their allies a way of browbeating them into accepting their assigned role in the new liberal economy. It is the flip side of redlining as a technique for twisting the arms of corporations: no one wants to be a bad corporate citizen any more than they want to be accused of institutional racism.

But the purpose of capitalism should be pure and simple. As Milton Friedman argued in his book *Capitalism and Freedom*, "There is one and only one social responsibility of business—to use its resources and engage in activities designed to increase its profits so long as it stays within the rules of the game, which is to say, engages in open and free competition, without deception and fraud." As Friedman puts it, businesses should not be immoral. But they should also not be subject to the moral whims of other people. Otherwise a situation is created in which political struggle, not financial prudence, defines what corporations do. In such a situation, the activist group with the biggest mouths or the most threatening thugs will rule the day.

Friedman goes on, "If businessmen do have a social responsibility other than making maximum profit for stockholders, how are they to know what it is? Can self-selected private individuals decide what the social interest is? Can they decide how great a burden they are justified in placing on themselves or their stockholders to serve that social interest?" Of course the activists don't want the businessmen to decide what their social responsibility might be; they want to decide it for them.[23]

But the problem lies not just with activists but with the Obama administration itself, which apparently does not see making money as a "private activity." As the investor Francis Cianfrocca points out, the Obama administration seems to believe that wealth creation "is only good if it can be used in furtherance of large-scale public goals." This was the same socialist ideology that lay behind the CRA.[24]

We should expect more financial crises in the future, starting in the very sector that triggered the collapse in the first place: real estate.

As noted above, the Obama administration has redoubled its commitment to aggressively enforcing the CRA while shoveling billions into the maw of ACORN and other activist organizations. Privatizing Fannie Mae and Freddie Mac would eliminate the risk of another mortgage crisis, but our old pal Barney Frank has balked at such a proposal because it would compromise affordable housing programs. California Democrat Joe Baca, the head of the Congressional Hispanic Caucus and a key proponent of extending cheap mortgage credit to Hispanics—a key Democratic constituency—likewise asserts, "We need to keep credit easily accessible to our minority communities." Instead of being reformed, the GSEs have been nationalized.

Meanwhile, the federal government has tried to jump-start the housing market through what Steven Malanga calls "ill-conceived tax credits and renewed federal subsidies for mortgages, including the Obama administration's mortgage bailout plan." As the Harvard economist Edward Glaeser observes, just as mortgage lenders have finally "recovered their sanity," the government is tempting them all over again with tax credits and low interest rates.[25] Far from stabilizing the housing market, such policies promise to pump it back up like a balloon ready to pop.

Nor is that the only bubble that the liberal state capitalists in the Obama administration are preparing for us. The CRA model of activist collusion with government to rig the capitalist system for approved social ends is now being applied in other areas.

It's not enough that we have the most advanced health care system in the world. Since it does not conform to its sense of social justice, the Obama administration is going to reengineer it, manipulate it, and alter it, creating a whole series of incentives and disincentives to bring it into line with the European health care model technocratic liberals admire. Thus we need a government-run insurance option to compete with private coverage. Never mind that by tinkering with the highly complex health care market they would undoubtedly make a hash of it. And what will happen if it blows up? Just as they blamed the mortgage crisis on unscrupulous lenders, Obama and his allies will blame the failed health care system on greedy HMOs and private doctors. After all, it is always someone else's fault.

The liberal Obamacrats also want to "fix" the auto industry. Never mind that none of the car mandarins in Washington has ever worked in the auto industry. The Obama administration deems that the solution to declining auto sales is to get Detroit to produce more fuel-efficient "green" cars. General Motors is going to make electric cars here, in the United States. Is there a market for them? Who knows? And who cares? As far as they are concerned, that is not really the point. There *should* be a market for these cars, and if we apply the right combination of carrot and stick people will buy them—at least they think they will. In the meantime, the Obama administration is pouring tens of billions of dollars of taxpayers' money into the auto companies in an effort to make them conform to their image of what car companies should be. Look for signs of an automotive bubble in the not-too-distant future.

But the capstone of liberal state capitalism in the age of Obama is the administration's effort to pump up the so-called green economy. Technological advances that create cleaner energy should be applauded, but the Obama people are engaging in precisely the sort of social engineering that housing advocates have been pushing for three decades. Forcing the creation of subsidized markets to promote expensive technologies for which there is no real demand distorts the economic system and creates investment bubbles.

Ignoring economic reality and the lessons of the current catastrophe, President Obama has pushed for "cap and trade," a scheme to create a greenhouse gas emissions trading market that will allow businesses to buy and sell the right to pollute. Never mind that there is no real-world value for carbon credits; Obama thinks a government mandate will make it so and create wealth in the process.

Most American corporations are not so foolish, but they do understand that if they can get on the inside they can profit from this arrangement. According to the Center for Public Integrity, more than two thousand lobbyists, hired by Wall Street and other business interests, pushed for the cap-and-trade scheme because they believe it could net some $2 trillion in paper value.[26] China and India, the developing industrial giants that generate much of the world's carbon dioxide, made a mockery of this plan by refusing to go along with it. Ordinary Americans, meanwhile, can look forward to skyrocketing energy prices as the costs of fighting global warming are passed on to the consumer.

President Obama recently pushed a large package of subsidies for renewable energy and declared that it would "create millions of additional jobs and entire new industries." In 2009, Congress passed a law that will allow big businesses to receive a federal grant to cover 30 percent of the cost of installing green energy. But these massive federal subsidies are necessary because the technologies in question are simply not viable at this point. Moreover, such governmental mandates only encourage foolish decision making that will be subsidized by the taxpayer.

For example, as *Forbes* magazine recently reported, a commercial printer installed a windmill on the top of his plant in New Haven, Connecticut. It didn't make economic sense; he wouldn't save money by doing so. But he did get taxpayers to pay for 83 percent of the turbine. As the magazine put it, "Suddenly it's not so much how sunny or windy a site is, but rather how much money is available." As a result, "Government money will be poured into renewable projects that won't produce much energy."[27]

President Obama believes that our economic salvation will come by creating green jobs. People will build wind turbines and solar panels, and work in carbon-trading markets. Meanwhile, Detroit—now a partially owned subsidiary of the federal government—will turn out federally mandated hybrid vehicles for which there is little demand. Needless to say, all of these initiatives will require some sort of taxpayer subsidy.

Do-good capitalism at work!

Perhaps no country in the world has spent more money and committed more resources to green energy than Spain. Yet a March 2009 report by Economics Professor Gabriel Calzada at Universidad Rey Juan Carlos found that the green jobs strategy has been disastrous for the country. The "green jobs" created by the government have been enormously expensive, requiring $752,000 to $800,000 in subsidies per job. Wind-power jobs cost $1.4 million each. These subsidies, of course, mean that other jobs will be lost. Calzada found that for every job created in the new state capitalist "green economy," at least 2.2 jobs are lost.[28]

The green-tech business model must include a government mandate that will force millions of other people to embrace the technology so they can have any hope of turning a profit. Jesse Ausubel, the director of the Program for the Human Environment at Rockefeller University, says flat out, "Alternative energy is the next subprime mortgage meltdown."

Perhaps the surest sign that we are headed for a green-tech bubble was the announcement by Heidi Fleiss, the notorious Hollywood Madam, that she was canceling plans to open a male brothel in Nevada and was instead going into green technology. "That's where the money is," she told a newspaper. "It's the wave of the future." [29]

Pursuing a liberal agenda is not a sustainable economic strategy. Using government regulatory power to create an artificial value does not create prosperity. It only creates the appearance of prosperity.

ACKNOWLEDGMENTS

A special heartfelt thanks goes to Adam Bellow, longtime editor and great friend, who not only fired the pistol to get me out of the starting gates but helped to drag me across the finish line in what was a 2½ month run to write and complete this book. This book would not have happened without you.

I have now celebrated ten years as a fellow at the Hoover Institution at Stanford University and remain convinced that the greatness of institution rests not in the vast resources it possesses but the people who work there. To John Raisian, the director, I am most appreciative for his support and friendship over the years. He manages to run the place with both wisdom and good cheer. I am also grateful to Noel Kolak, Stephen Langlois, and Richard Sousa who are constantly making my job easier, as well as Deborah Ventura, Celeste Szeto, and Nancy Cloud.

On this particular book I was blessed to have the support and assistance of a team that assisted greatly in producing this book on such a short schedule. Rhonda Adair not only helped with shaping the manuscript and organizing the team but also offered many creative suggestions and important intellectual input to the project. (Thanks also go to Ava and Raquel for being so patient during those long meetings.) Jonathan Nicholson, fresh out of a college, proved to be a most diligent researcher, who brought important suggestions and insights with him every week to our meetings. Peggy Sukhia proved equally effective in tracking down facts and details that somehow I lost or couldn't find in the first place. Thanks Jonathan and Peggy. You both have great futures ahead of you. And what can I say about Rick White? A long-time friend and a former top securities regulator in the State of Florida, he was vital in helping me to master the intricacies of financial regulations and the institutions at play in this story. He was generous with his time even though he was launching his new venture, Turris Consulting, as I bothered him with all sorts of crazy questions. Rick also offered great editorial advice.

Thanks, too, go to Brian Baugus, PhD, who was also kind enough to read through the manuscript and offer his comments. And I also benefitted from the friendship and thoughts of Wynton Hall (who suggested the title), Ron Robinson, Michelle Easton, David and Amy Ridenour, Hugh Nicholson, Paul Kengor, Randy Enwright, Steven Bannon, Steve Post, Tim Ireland, Tom Sylte, Jeremy Cohen, John Bickley, and Stuart Christmas.

Marc Thiessen, a good friend and my partner at Oval Office Writers Group, is a constant source of encouragement and inspiration. Thank you my friend.

I'm grateful to Glen Hartley and Lyn Chu, my new agents, for their diligent work. Thanks also go to my long-time publicist Sandy Schulz and her husband Max for their friendship.

Completion of this book took place at a particularly difficult personal time for me and I'm grateful for our dear friends Richard and Elizabeth Albertson, David and Becky Healy, Bill and Jill Mattox, Bob and Debbie Evans, Jimmy and Karen Hill, and Anthony and Sally Jo Roorda, for their friendship.

Thanks finally goes to my family. You have put up with crazy hours, distractions, and stress as I worked to complete this manuscript. My beautiful wife Rochelle, to whom this book is dedicated, has been through many book projects with me, but this one was different as she dealt with a variety of family challenges. Thanks for working through tough times. My son Jack, 12-years-old and full of life, is a great encourager and motivator. Jack, I love to see you grow up. I'm going to be so excited to see what you do. Hannah, AKA as "Sweet Pea," you manage to mix grace and spunk together all in one. Thanks for being you.

I also want to thank my mother, Kerstin Schweizer, for her continued love, encouragement and support, as well as my sis Maria Duffus, her husband Joe, nephews Adam and Danny, and my brother-in-law Richard Rueb.

As always, the author alone is responsible for the contents of his book.

NOTES

INTRODUCTION: A FAILURE OF CAPITALISM—OR LIBERAL SOCIAL ENGINEERING?

1 Jeff Jacoby, "Frank's Fingerprints Are All over the Financial Fiasco," *Boston Globe*, September 28, 2008.

2 Kiyoshi Okonogi, "Paul Samuelson: Financial Crisis Work of 'Fiendish Monsters,' " Asahi Shimbun, October 31, 2008; see also "Paul Samuelson," *New Perspectives Quarterly*, January 16, 2009.

3 Lawrence Summers, "The Pendulum Swings Toward Regulation," *Financial Times*, October 26, 2008.

4 Scott S. Powell, "The Culprit Is All of Us," *Barron's*, February 9, 2009.

5 See John Taylor, *Getting off Track: How Government Actions and Interventions Caused, Prolonged, and Worsened the Financial Crisis* (City?: Hoover Institution Press, 2009).

6 Ellen McGirt, "Al Gore's $100 Million Makeover," Fast Company, Feb. 8, 2008.

7 Kara Rowland, "Rep. Perlmutter Part Owns 'Green' Bank He Helped," *Washington Times*, July 15, 2009.

CHAPTER 1: THE ROBIN HOOD AGENDA: HOW A GANG OF RADICAL ACTIVISTS AND LIBERAL POLITICIANS SET THE STAGE FOR THE BIGGEST BANK HEIST IN HISTORY

1 *Selma S. Buycks-Roberson v. Citibank Federal Savings Bank*, Complaint no. 94C 4094. United States District Court, Northern District of Illinois.

2 Ibid.

3 *Selma S. Buycks-Roberson v. Citibank Federal Savings Bank*, no. 94C 4094, transcript of proceedings before the Honorable Ruben Castillo, United States District Court, Northern District of Illinois, November 12, 1997, 5.

4 See, e.g., Abdon M. Pallasch, "Obama's Legal Career: He Was 'Smart, Innovative, Relentless,' and He Mostly Let Other Lawyers Do the Talking," *Chicago Sun-Times*, December 17, 2007.

5 *Selman S. Buycks-Robersons v. Citibank Federal Savings Bank*, Settlement Agreement No. 94 C 4094 United States District Court, Northern District of Illinois.

6 Debra Kroupa, "Community Reinvestment Act Has Worked," *Dallas Morning News*, May 31, 1999.

7 Heidi Swarts, *Organizing Urban America: Secular and Faith-based Progressive Movements* (University of Minnesota Press, 2008), 77.

8 Howard Husock, *America's Trillion-Dollar Housing Mistake: The Failure of American Housing Policy* (Ivan R. Dee, 2003), 64.

9 Saul Alinsky, *Rules for Radicals,* (Vintage, 1989).

10 Ryan Lizza, "The Agitator," The New Republic, March 19, 2007.

11 Pete Harrison, "Never Waste a Good crisis," Reuters, March 7, 2009.

12 Barack Obama, *After Alinsky: Community Organizing in Illinois* (University of Illinois at Springfield, 1990).

13 Quoted in George Beam, *Strategies for Change: How to Make the American Political Dream Work* (Swallow Press, 1976), 128.

14 Douglas Martin, "Gale Cincotta, 72, Opponent of Biased Banking Policies," *New York Times,* 9/17/01.

15 "HUD Aide's Home Is Target of Protest 2nd Night; Arrest 2," *Chicago Tribune*, April 13, 1972; Gerald Conner, "Group Says It May Picket Daley Home," *Chicago Tribune*, December 22, 1970; Tom Slocum, "Austin Organization Repudiates Charges of Badgering Businesses," *Chicago Tribune*, July 30, 1970.

16 Dan Immergluck, *Credit to the Community* (M. E. Sharpe, 2003), 141–142.

17 Terry H. Anderson, *The Movement and the Sixties* (New York: Oxford University Press, 1996), 387.

18 Richard Marsico, *Democratizing Capital, The History, Law, and Reform of the Community Reinvestment Act* (Carolina Academic Press, 2005), 3.

19 Dr. Harry Edwards, "The Man Who Would Be King in the Sports Arena," ESPN.com, Feb. 28, 2002.

20 "Jackson to Ask U.N. for Cash to Help Finance Rights Group," *Chicago Tribune*, December 25, 1971.

21 "Operation Breadbasket," *Chicago Tribune*, December 10, 1966; and Alan Merridew, "Financial Leaders to support PUSH," *Chicago Tribune*, November 18, 1973.

22 Steve Miller, "Targets of Boycott Gave to Jackson: Tax Forms Reveal Revenue Sources," *Washington Times*, April 17, 2002.

23 "Jesse Jackson Raps Japanese on Redlining," *Charlotte Observer*, December 9, 1986.

24 Jo Ann S. Barefoot, "Has CRA Become Anti-bank Activists' New All-Purpose Tool?," *ABA Journal* 90, no. 8 (1998).

25 Steven Malanga, "Obsessive Housing Disorder," *City Journal*, Spring 2009, 20.

26 Mara S. Sidney, *Unfair Housing: How National Policy Shapes Community Action* (University Press of Kansas, 2003), 45.

27 Quoted in Marsico, *Democratizing Capital*, 13.

28 *Congressional Record*, 123: 17630.

29 Quoted in Sidney, *Unfair Housing*, 44.

30 Ibid., 40–44.

31 Sidney, 46.

32 Quoted in Marsico, *Democratizing Capital*, 15.

33 *Congressional Record*, 123 (date?): 17,628 and 17,636.

34 Judith Miller, " 'Community' Law on Loans Debated," *New York Times*, April 24, 1978; "Maximizing Mortgages, Minimizing Risks," *New York Times*, November 27, 1977.

35 See, e.g., Dan Immergluck, *Credit to the Community: Community Reinvestment and Fair Lending Policy in the United States* (M. E. Sharpe, 2004), 106; Barefoot, "Has CRA Become Anti-bank Activists' New All-Purpose Tool?"

36 Susan Hoffman, *Politics and Banking: Ideas, Public Policy, and the Creation of Financial Institutions* (Baltimore, Md.: Johns Hopkins University Press, 2001).

37 Heidi J. Swarts, *Organizing Urban America* (University of Minnesota Press, 2008), 86.

38 Ibid., 31.

39 Maude Hurt, national president of ACORN, letter to the *Washington Times*, January 25, 1999.

40 *To Each Their Home: Success Stories from the ACORN Housing Corporation*, (brochure).

41 Swarts, *Organizing Urban America*, 86.

42 Matthew Vadum, "ACORN: Who Funds the Weather Underground's Little Brother?," Capital Research Center Foundation Watch, City?, November 2008.

43 Stephanie Strom, "Funds Misappropriated at 2 Non-profit Groups," *New York Times*, July 9, 2008.

44 Swarts, *Organizing Urban America*, 10.

45 National Community Reinvestment Coalition, *CRA Commitments* [a report], September 2007, 4.

CHAPTER 2: $4 TRILLION SHAKEDOWN:
THE LEFT'S ACTIVIST JIHAD AGAINST AMERICAN BANKS

1 David Everett, "Confrontation, Negotiation, and Collaboration: Detroit's Multibillion Dollar Deal," in *From Redlining to Reinvestment: Community Responses to Urban Disinvestment*, ed. Gregory Squires (Philadelphia: Temple University Press, 1992), 109.

2 John McCarron, "Quiet Persuaders Use Law to Build Up Inner City," *Chicago Tribune*, January 16, 1985.

3 Carol Memmott, "Activist Fights for Housing," *USA Today*, April 10, 1989.

4 Neal Peirce, "Community Groups Instill Fear in Banking Circles More Dollars Flowing into Troubled Neighborhoods Thanks to Community Reinvestment Act," *Charlotte Observer*, July 18, 1987.

5 Kevin Haney and Marc Melter, "Putting the Squeeze on Banks, Activists Use New Tactics to Get Loans for Poor," *Philadelphia Daily News*, October 7, 1986.

6 Vivenne Walt, "Chase Move Is Disputed," *Newsday*, July 3, 1989.

7 Elizabeth Sanger, "Groups Hit Manny Hanny," Newsday, May 5, 1989 .

8 Susan Harrigan, "Civic Group Gets Bank's Promise for Below-Market Loans in Bronx," *Newsday*, November 6, 1987.

9 "Another ACORN Scandal," *New York Post*, July 13, 2008; Steven Erlanger,

"New York Turns Squatters into Homeowners," *New York Times*, October 12, 1987.

10   Bob Liff, "Bank Deal to Include $10 Million for Housing," *Newsday*, March 4, 1988.

11   Christian Wihtol, "Carteret to Help Poor with Loans," *The Record* (Bergen County, N.J.), February 3, 1988.

12   Kenneth Tompkins, "Reacting to Pressure—Bankers Say They're Doing Better Job Meeting Credit Needs," *Post-Standard* (Syracuse, N.Y.), April 29, 1988.

13   Joanne Ball, "Bank Accused of Bias in Granting Mortgages," *Boston Globe*, February 1, 1986.

14   Bruce A. Mohl, "Bank to Reserve Loans for Low-Income Areas," *Boston Globe*, February 23, 1980.

15   David Warsh, "Bank Commissioner Denies Provident Branch in Newton," *Boston Globe*, January 3, 1980.

16   Sean Murphy, "Somerville Bank Pledges $25 Million in Housing Loans," *Boston Globe*, February 10, 1988.

17   Dan Immergluck, *Credit to the Community: Community Reinvestment and Fair Lending Policy in the United States* (M.E. Sharpe, 2004) op cit., Au: Full ref., please 165.

18   Stanley Ziemba, "Redlining Fight Bearing Fruit in Money-Strapped Inner Cities," *Chicago Tribune*, May 25, 1986.

19   Hank De Zutter, "What Makes Obama Run?," *Washington City Paper*, January 22, 2009. This article originally appeared in the *Chicago Reader* on December 8, 1995.

20   Jay McIntosh, "Activists Set Deadline for Bank to Aid Loan Fund," *Charlotte Observer*, June 24, 1989.

21   Steve Matthews, "The Key to Bank Profitability? It All Depends on the Measure," *Charlotte Observer*, October 8, 1986; James Mallory, "Serve All of Community, Bankers Told," *Atlanta-Journal Constitution*, May 5, 1989.

22   David F. Mildenberg, "First Union to Expand Minority, Low-Income Programs," *Charlotte Observer*, July 4, 1986.

23   Nancy L. Ross, "Pratt Says Regulations of S&Ls Shouldn't Dictate Loan Choices," *Washington Post*, July 2, 1981.

24   Glenn Canner, "The Community Reinvestment Act and Credit Allocation,"

Unpublished report, Washington, D.C., Federal Reserve System. Quoted in Dan Immergluck, *Credit to the Community*, op cit. p 159.

25 Douglas Demmons, "Hibernia Claims It Won Dispute," *The Advocate*, July 26, 1986, and "Citizens Forcing Banks to Change," *The Advocate*, May 15, 1986.

26 Andrea Knox, "Continental Bank Plans to Branch Out in Low-Income Lending," *Philadelphia Inquirer*, August 26, 1986; Neal Peirce, "Community Groups Instill Fear in Banking Circles; More Dollars Flowing into Troubled Neighborhoods Thanks to Community Reinvestment Act," *Charlotte Observer*, July 18, 1987.

27 Jim Erickson, "More Protests Question Lending Practices of Banks," *Seattle Post-Intelligencer*, January 20, 1988.

28 David Satterfield, "Banks Prodded to Serve Poor," *Miami Herald*, September 15, 1985.

29 Goldie Blumenstyke, "Orlando's Poor Lose Advocate Jay Rose," *Orlando Sentinel*, November 2, 1986.

30 Goldie Blumenstyke, "Bank's Promise Halts Complaint," *Orlando Sentinel*, July 19, 1985.

31 Arnold Markowitz, "Jury: Banks Should Aid Black Firms," *Miami Herald*, February 7, 1985.

32 Tim Smart, "Barnett Banks Affirms Loan Commitment," *Orlando Sentinel*, September 9, 1986.

33 Amy Wallace, "Trust Co. Plan for New Branch in Cherokee County Draws Fire," *Atlanta Journal-Constitution*, May 26, 1989.

34 Kenneth R. Harney, "Using the Community Reinvestment Act," *Washington Post*, October 6, 1990.

35 Scott Clark, "Banks Step Up Lending to Low-Income Areas—Civic Groups Pressure Lenders Under Reinvestment Act," *Houston Chronicle*, September 28, 1987.

36 David Nicklaus, "ACORN, Centerre Agree," *St. Louis Post-Dispatch*, August 31, 1988.

37 John Ikeda, "Action Smoothes Merger: Cal First to Boost Minority Lending," *Evening Tribune*, June 24, 1988.

38 Susan Chandler, "Banks Paint Two Pictures of Roseland Activist Lomax," *Chicago Sun-Times*, July 24, 1989; Susan Chandler, "Battling the Color of Money—Black Banks Squeezed by Community Needs, Wants," *Chicago Sun-Times*, July 23, 1989.

39 Hiawatha Bray, "Critic Calls Low-Income Lending Law Illegal," *Detroit Free Press*, August 31, 1993.

40 Mary Sit, "Banks with 'White Hats' Help N.E.'s Poor," *Boston Globe*, March 20, 1989.

41 Associated Press, "Home Loan Practice Decried—Joe Kennedy Cites Bias Nationwide," *Austin American-Statesman*, January 24, 1989.

42 David Everett John Gallagher, and Teresa Blossom, "The Race for Money," *Detroit Free Press*, July 24–27, 1988.

43 David Everett, "Confrontation, Negotiation, and Collaboration: Detroit's Multibillion-dollar deal," in Gregory D. Squires, *From Redlining to Reinvestment: Community Responses to Urban Disinvestment* (Temple University Press, 1992) p 120.

44 David Everett, Teresa Blossom, and Tim Gallagher, "Proxmire Urges Probe of Bank Loan Pattern," *Detroit Free Press*, July 27, 1988.

45 David Goldberg, "Discrimination in Mortgage Lending: An Examination of Several Data Sets Pertaining to Comerica, 1986," quoted in Everett, "Confrontation, Negotiation, and Collaboration: Detroit's Multibillion-dollar deal," op cit. p 124–125.

46 Ibid., 125.

47 Jeffrey Gunther, "Should CRA Stand for Community Redundacy Act?," *Regulation* 23, no. 2.

48 Kathleen C. Engel and Patricia McCoy, "The CRA Implications of Predatory Lending," *Fordham Urban Law Journal* 29, 2002.

49 National Community Reinvestment Coalition, *CRA Commitments*, September 2007, 33.

40 Peter J. Wallison, "Cause and Effect: Government Policies and the Financial Crisis," *Critical Review* 21, nos. 2–3.

51 *Wall Street Journal*, February 20, 2009.

52 Edward Gramlich, Remarks at the Financial Services Roundtable Annual Housing Policy Meeting, May 21, 2004.

CHAPTER 3: THE CLINTON CRUSADE: HOW DEMOCRATS MADE CREDIT A CIVIL RIGHT

1 Dan Immergluck, *Credit to the Community* (M. E. Sharpe, 2003), 174.

2 Peter Slevin, "For Clinton and Obama, a Common Ideological Touchstone," *Washington Post*, March 25, 2007.

3 See Robert Rubin and Jacob Weisberg, *In an Uncertain World: Tough Choices from Wall Street to Washington* (New York: Random House, 2003), Au: Page?.

4 Loic Sadoulet, "Microcredit Repayment Insurance: Better for the Poor, Better for the Institution," in *Credit Markets for the Poor*, ed. Patrick Bolton and Howard Rosenthal (Russell Sage Foundation, 2005), 199–200.

5 John H. Makin, "A Government Failure, Not a Market Failure," *Commentary*, July-August 2009, 18.

6 Paul Anderson and Janet Reno, *Doing the Right Thing* (New York: John Wiley, 1994).

7 Pierre Thomas, "Justice Department Civil Rights Chief Pledges Activism," *Washington Post*, April 15, 1994.

8 Brian C. Mooney, "Patrick's Path from Courtroom to Boardroom," *Boston Globe*, August 13, 2006.

9 Frank Phillips, "Governor Made Call on Behalf of Lender—Troubled Ameriquest—Sought Infusion of Cash," *Boston Globe*, March 6, 2007.

10 Bruce Bartlett, "Roberta on a Rampage," *National Review*, May 2, 1994.

11 Angela D. Chatman, "Fair-Housing Optimism Is Infectious at Summit," *Cleveland Plain Dealer*, January 30, 1994.

12 "There's No 'White's Only' Sign, But . . ." *BusinessWeek*, Oct. 26, 1992.

13 "Cisneros Pledges Efforts to End Racial Disparity," *Mortgage Banking*, November 1993.

14 See George J. Benston, "The Community Reinvestment Act: Looking for Discrimination That Isn't There," *CATO Institute Policy Analysis*, October 6, 1999; David K. Horne, "Mortgage Lending, Race, and Model Specification," *Journal of Financial Services Research* 11, nos. 1–2 (1997): 43–68; Harold Black, "Discrimination in Financial Services," *Journal of Financial Services Research* 11, nos. 1–2 (1997): 189–204.

15 Becker, quoted in Carolyn M. Brown and Matthew S. Scott, "How to Fight Mortgage Discrimination . . . and Win!!! African Americans Join Forces to End Racist Lending Practices. Black Enterprise Reviews Fight-Back Techniques That Can Work for You," *Black Enterprise*, July 1993. See also Dinesh D'Souza, *The End of Racism* (New York: Free Press, 1995), 279–282.

16 Robert Stowe England, "Washington's New Numbers Game," *Mortgage Banking*, September 1993.

17 Associated Press, "Clinton Attacks Housing Bias Power of Courtroom Persuasion Will Produce Results, Reno Says," *Roanoke Times*, November 5, 1993.

18 Paul Anderson and Janet Reno, op cit. p 86.

19 John H. Cushman, Jr., "Fight Against Bias Difficult for Banks," *Times Union* (Albany, N.Y.), November 28, 1993.

20 Steve Cocheo, "Justice Department Sues Tiny South Dakota Bank for Loan Bias," *ABA Banking Journal*, January 1, 1994.

21 Tony Pugh, "New Edict on Redlining Rocks Banks," *Miami Herald*, August 28, 1994.

22 Jerry Seper, "Bank Offers Cut-Rate Mortgages to End Redlining Probe," *Washington Times*, August 14, 1997.

23 Steve Cocheo, "Fair-Lending Pressure Builds," *ABA Banking Journal*, December 1, 1994.

24 John R. Walter, "The Fair Lending Laws and Their Enforcement," *Federal Reserve Bank of Richmond Economic Quarterly* 81, no. 4.

25 Brown and Scott, "How to Fight Mortgage Discrimination . . . and Win!!!"

26 Associated Press, "U.S. Expands Lending Discrimination Definition," *Arizona Daily Star*, March 9, 1994; Dennis Sewell, "Clinton Democrats Are to Blame for the Credit Crunch," *The Spectator* (U.K.), October 1, 2008.

27 Roger Fillion, "U.S. Unveils Lending Guidelines," *Pittsburgh Post-Gazette*, March 9, 1994.

28 Lawrence B. Lindsey, "A Balanced Response," *Mortgage Banking*, October 1993.

29 Robert Stowe England, "Washington's New Numbers Game," *Mortgage Banking*, September 1993.

30 Eugene A. Ludwig, "The Quiet Revolution," *Mortgage Banking*, September 1997.

31 England, "Washington's New Numbers Game."

32 Nicholas M. Horrock, "U.S. Vows to Redouble Fair-Lending Fight," *Chicago Tribune*, November 5, 1993.

33 Associated Press, "Reno Says Bill Will Turn Back the Clock," *The State* (Columbia, S.C.), June 16, 1995.

34 Steve Massey, "Lincoln Savings' Loans Probed; Community Group Says Institution Discriminates Against Black People," *Pittsburgh Post-Gazette*, December 22, 1993.

35 Steve Cocheo, "Washington Turns Fair-Lending Volume to MAX," *ABA Banking Journal* 86, no. 3, 1994.

36 Ibid.

37 Ibid.

38 Phillip L. Schulman, "No Paper Tiger," *Mortgage Banking*, October 1993.

39 David E. Bernstein, *You Can't Say That! The Growing Threat to Civil Liberties from Antidiscrimination Laws* (Cato Institute, 2003), 77.

40 Quoted in Peter J. Wallison, "Cause and Effect: Government Policies and the Financial Crisis," *Critical Review* 21, nos. 2–3: 366.

41 David Streitfeld and Gretchen Morgenson, "Building Flawed American Dreams," *New York Times*, October 19, 2008.

42 Eugene A. Ludwig, "The Quiet Revolution," *Mortgage Banking*, September 1997.

43 Wayne Barrett, "Andrew Cuomo and Fannie and Freddie," *Village Voice*, Aug. 5, 2008.

44 Patrice Hill, "Easy Credit Turning into Hard Times?," *Insight on the News*, November 8, 1999.

45 Alan J. Heavens, "Building Paths Past Home Loan Prejudices: Redlining Has Long Been Illegal, but Discrimination Takes Subtle Forms," *Philadelphia Inquirer*, February 13, 1994.

46 *Congressional Record*, (May 24, 1995), E116; and (October 26, 1995), E2057.

47 See Stanley Kurtz, "Spreading the Virus," *New York Post*, October 14, 2008; Matthew Vadum, "ACORN: Who Funds the Weather Underground's Little Brother?" Capital Research Center Foundation Watch, November 2008.

48 Benjamin Dattner, "A Powerful Partnership," *Mortgage Banking*, October 1993.

49 National Community Reinvestment Coalition, *CRA Commitments*, September 2007.

50 Ned Brown and Dale Westhoff, "Packaging CRA Loans into Securities," *Mortgage Banking*, May 1998.

51 National Community Reinvestment Coalition, *CRA Commitments*, September 2007.

52 Ellen Seidman, "CRA in the 21st Century," *Mortgage Banking*, October 1999.

53 Roger Fillion, "Bankers Say Clinton Uses Them to Push a Costly Social Agenda," *Pittsburgh Post-Gazette*, February 18, 1994.

54 Eric S. Belsky, Michael Schill, and Anthony Yezer, "The Effect of the Community Reinvestment Act on Bank and Thrift Home Purchase Mortgage Lending," Harvard University Joint Center for Housing Studies, 2001.

55 Sabrina Tyuse and Julie Birkenmaier, "Promoting Homeownership for the Poor: Proceed with Caution," *Race, Gender and Class* 13, no. 3–4: 295.

56 Ibid.; Timothy Canova, "The Clinton Bubble," *Dissent*, Summer 2008.

57 Remarks of the Honorable Janet Reno, Attorney General of the United States, to the National Community Reinvestment Coalition, March 20, 1998.

58 Robert Rubin, keynote address, "Money, Markets and the News: Press Coverage of the Modern Revolution in Financial Institutions," Joan Suorenstein Center, Harvard University, March 1999, p. 14.

59 Rubin and Weisberg, *In an Uncertain World*, 200.

60 *Congressional Record* (October 26, 1995), E2057.

61 Beth Healy, "Rubin Eyes Economy at Urban Level," *Boston Globe*, June 20, 2001.

62 Raisa Bahchieva, Susan M. Wachter, and Elizabeth Warren, "Mortgage Debt, Bankruptcy, and the Sustainability of Homeownership," in Patrick Bolton and Howard Rosenthal, *Credit Markets for the Poor* (Russell Sage Foundation, 2005), 200.

63 Raisa Bahchieva, op cit.

64 Brown and Westhoff, "Packaging CRA Loans into Securities."

65 Ellen Seidman, "CRA in the 21st Century," *Mortgage Banking*, Oct. 1999.

CHAPTER 4: COVER YOUR FANNIE: HOW FANNIE MAE AND FREDDIE MAC WERE TAKEN OVER BY LIBERAL ACTIVISTS

1 "NTIC celebrates 30 years of organizing neighborhoods," *Disclosure*, October 31, 2003.

2 *Congressional Record* (June 20, 2000), H4767.

3 "The Cycle of Organizing: Revitalizing Old Campaigns with New Strategies," *Disclosure*, October 31, 2002.

4 Ben Wiederholt, "CIOP Mortgage Agreement Brightens Holidays for Families," *Disclosure*, December 31, 2003.

5 "Mini-NPA Sets Up National Conference in June," *Disclosure*, April 30, 2002.

6 Peter Dreier, "Why America's Workers Can't Pay the Rent," *Dissent*, Summer 2000.

7 Mike Montag, "Mortgages Aim to Inhibit Sprawl," *World Watch*, November–December 2000.

8 See the IRS 990-PF forms filed by the Fannie Mae Foundation and Freddie Mac Foundations.

9 Paul Muolo and Mathew Padilla, *Chain of Blame* (New York: Wiley, 2008), 111.

10 Richard W. Stevenson, "The Velvet Fist of Fannie Mae," *New York Times*, April 20, 1997.

11 Janet Hewitt, "Credit Scoring, Credit Quality and Lobbying," *Mortgage Banking*, Nov. 1, 1995.

12 Ron Brownstein, "Behind the Boom in Minority Homeownership," *Los Angeles Times*, June 13, 1999.

13 Carol D. Leonnig, "How HUD Mortgage Policy Fed the Crisis," *Washington Post*, June 10, 2008.

14 "Fannie Mae Moves to Aid Affordable Loans," *Washington Times*, October 25, 2006.

15 Natasha Shulman, *Reaching the Immigrant Market: Creating Homeownership Opportunities for New Americans.* Institute for the Study of International Migration, Georgetown University, 2003, p 56.

16 Ibid., 54.

17 Kenneth Temkin, George Galster, Roberto Quercia, and Sheila O'Leary, "A Study of the GSEs' Single-Family Underwriting Guidelines," Executive Summary, Urban Institute, Washington, D.C., April 9, 1999.

18 Martin Luther King III, "Minority Housing Gap—Fannie Mae, Freddie Mac Fall Short," *Washington Times*, November 17, 1999.

19 Associated Press, "HUD Raises Requirement of Loans for 2 Financers," July 30, 1999.

20 Wayne Barrett, "And now Cuomo and Fannie and Freddie," *Village Voice*, Aug. 5, 2008.

21 Jesse Jackson and Jesse L. Jackson, Jr., *It's About the Money!* (Three Rivers Press, 1999.)

22 Cindy Loose, "Racial Disparity Found in Credit Rating," *Washington Post*, Sept. 21, 1999.

23 Carol Leonnig, "How HUD Mortgage Policy Fed the Crisis," *Washington Post*, June 10, 2008.

24 Peter Wallison, "Cause and Effect: Government Policies and the Financial Crisis," *Critical Review* 21, nos. 2–3 2009; Wayne Barrett, "Andrew Cuomo and Fannie and Freddie," *Village Voice*, August 5, 2008.

25 "Fannie Mae Increases CRA Options," ABA Banking Journal 92, no. 11 (2000).

26 Richard Koonce, "Redefining Diversity: It's Not Just the Right Thing to Do," *Training and Development* 55 (December 2001).

27 Office of Federal Housing Enterprise Oversight, Report on Findings of the Special Examiniation of Fannie Mae, Sept. 17, 2004.

28 "Fannie Mae Bending Financial System to Create Homeowners," says Raines, Univ. of Connecticut *Advance,* Oct. 30, 2000.

29 Steven A. Holmes, "Fannie Mae Eases Credit to Aid Mortage Lending," *NY Times* Sept. 20, 1999.

30 Muolo and Padilla, *Chain of Blame*, 113.

31 Quoted in ibid.

32 Quoted in ibid., 256.

33 "American Dream Builder," *NYSE* [New York Stock Exchange] *Magazine*, May 2005.

34 Muolo and Padilla, *Chain of Blame*, 116.

35 Quoted in ibid., 289.

36 Rodolfo Saenz, "The Emerging-Market Opportunity," *Mortgage Banking*, May 2004.

37 "Excerpts of emails from Angelo Mozilo," SEC.GOV/NEWS/PRESS/2009/2009_129_EMAIL.HTM.

38 "Mortgage Maker vs. the World," *New York Times*, October 16, 2005.

39 Muolo and Padilla, *Chain of Blame*, 262.

40 Susan Schmidt and Maurice Tamman, "Housing Push for Hispanics Spawns Wave of Foreclosures," *Wall Street Journal*, January 5, 2009.

41 Daniel Golden, "Countrywide's many 'Friends,'" Portfolio, June 12, 2008.

42 Patrick McGreevy, "Mayor Is Criticized for Jet Use," *Los Angeles Times*, November 8, 2005.

43 "Washington Mutual Wins 2003 CRA Community Impact Award," *CRSwire*, November 4, 2003; Robert O'Connor, "That's Affordable: Seattle-Based

Washington Mutual Has Built a Serious Business Around Community Outreach and Affordable Lending," *Mortgage Banking*, October 1, 2003.

44 Gretchen Morgenson and Geraldine Fabrikant, "Countrywide's Chief Salesman and Defender," *New York Times*, November 11, 2007.

CHAPTER 5: THE GOLDEN TROUGH: HOW LIBERAL POLITICIANS USED FANNIE AND FREDDIE TO RIG THE REAL ESTATE MARKET WHILE LINING THEIR OWN POCKETS

1 William Poole, "Housing in the Microeconomy," *Federal Reserve Bank of St. Lewis Review,* May/June 2003.

2 "Fannie Mae Liberals," *Wall Street Journal*, October 14, 2004.

3 Richard W. Stevenson, "The Velvet Fist of Fannie Mae," *New York Times*, April 20, 1997.

4 Bob Secter and Andrew Zajac, "Rahm Emanuel's Profitable Sting at Mortgage Giant," *Chicago Tribune*, March 26, 2009.

5 Gerald O'Driscoll, "Fannie/Freddie Bailout Baloney," *New York Post*, September 9, 2008.

6 "Christmas for Fannie Mae," *Wall Street Journal*, December 23, 2003; "Latin Lawmakers Urge Fed to Hold Fannie, Freddie Study," *Wall Street Journal*, December 19, 2003.

7 Owen Ullmann, "Crony Capitalism: American Style," *International Economy*, July–August 1999, 10.

8 Ibid.

9 "Frantic Fannie," *Wall Street Journal*, February 28, 2002.

10 Richard W. Stevenson, "The Velvet Fist of Fannie Mae," *New York Times*, April 20, 1997.

11 Jonathan G. S. Koppell, "Hybrid Organizations and the Alignment of Interests: The Case of Fannie Mae and Freddie Mac," *Public Administration Review,* July–August 2001.

12 Ullmann, "Crony Capitalism," 11.

13 "Fan and Fred Get the Business," "Fannie's Inside Info," *Wall Street Journal*, July 1, 2002

14 Office of Federal Housing Enterprise Oversight, Report on the Findings of the special examination of Fannie Mae, Sept. 17, 2004.

15 Gregory Mankiw, "Keeping Fannie and Freddie's House in Order," *Financial Times*, Feb. 24, 2004.

16 "White House Fannie Pack," *Wall Street Journal*, November 11, 2003.

17 Bob Secter and Andrew Zajac, "Rahm Emanuel's Profitable Stint at Mortgage Giant," *Chicago Tribune*, March 26, 2009.

18 "Fannie Mae's Risky Business," *Wall Street Journal*, September 23, 2002.

19 Alan Greenspan, Testimony before the Committee on Banking, Housing, and Urban Affairs, U.S. Senate, Feb. 24 2004.

20 "Fannie Mayhem," *Wall Street Journal*, November 20, 2007.

21 Letter from Alan Greenspan to the Honorable Robert F. Bennett, U.S. Senate, September 2, 2005.

22 September 2003 hearing of the House Committee on Financial Services.

23 Abramoff was an American lobbyist who is now serving in federal prison for committing fraud and corrupting public officials. The extensive corruption investigation into his activities led to the conviction of White House officials, a congressman, and congressional aides.

24 Michael Crittenden, "Sen. Dodd calls Fannie, Freddie 'Fundamentally Strong,'" Marketwatch, July 11, 2008.

25 Ullmann, "Crony Capitalism," 11.

26 Office of Federal Housing Enterprise Oversight, Report on the Findings of the Special Examination of Fannie Mae, Sept. 17, 2004.

27 "Memo to Fannie," *Wall Street Journal*, June 14, 2006.

28 Bethany McLean, "The Fall of Fannie Mae," Fortune Magazine, Jan. 24, 2005.

29 Rob Blackwell, "Fannie and OFHEO Take Lumps in House Hearing," *American Banker*, October 7, 2004.

30 "Rep. Frank Backs Away from OFHEO $," *National Mortgage News*, November 15, 2004.

31 "Government Sponsored Enron: Billion-Dollar Scandal Not Ready for Prime Time," Media Research Center News Analysis, March 22, 2005.

32 "Fannie Uncovered," *Wall Street Journal*, September 23, 2004.

33 "Too Political to Fail," *Wall Street Journal*, April 21, 2008.

34 Peter Schroeder, "Schumer Offers Bill Easing Caps on Fannie Mae and Freddie Mac," *Bond Buyer*, September 11, 2007.

35 "Fannie Mae Ugly," *Wall Street Journal*, July 12, 2008.

36 Vernon C. Smith, "The Clinton Housing Bubble," *Wall Street Journal*, Dec. 18, 2007.

CHAPTER 6: DO-GOOD CAPITALISTS: BILL CLINTON'S SEDUCTION OF WALL STREET AND THE BIRTH OF THE BAILOUT CULTURE

1   Gordon Matthews, "For a New Wall Street Generation, a Bear Is Just a Fuzzy Critter," *American Banker*, September 8, 1998.

2   Steve Fraser, *Every Man a Speculator: A History of Wall Street in American Life* (New York: Harper, 2005).

3   Brett Fromson, "Farm Boy to Financier—Goldman's Corzine Embodies Wall Street's New Meritocracy," *Washington Post*, November 6, 1994.

4   Gary Stix, "A Calculus of Risk," *Scientific American*, May 1998; Ludwig von Mises, *Human Action: A Treatise on Economics* (Yale University Press, 1949); Kevin Phillips, *Bad Money* (Penguin Books, 2009).

5   Kevin Phillips, op. cit., 183.

6   Teresa Lindeman, "As the New Economy Turns: Is There Any of It Left?" *Pittsburgh Post-Gazette*, April 8, 2003.

7   Daniel Gross, *Bull Run: Wall Street, the Democrats, and the New Politics of Personal Finance* (City?: Public Affairs, 2000)

8   Robert Rubin and Michael Rubinger, "Don't Let Banks Turn Their Backs on the Poor," *NY Times*, Dec. 4, 2004.

9   Philipps, *Bad Money* 172.

10 Steve Miller, "Targets of Boycott Gave to Jackson; Tax Forms Reveal Revenue Sources," *Washington Times*, April 17, 2002.

11 See for example David Hogberg and Sarah Haney, "Funding Liberalism with Blue-Chip Profits: Fortune 100 companies Bank Leftist Causes," *Foundation Watch*, Capital Research Center, August 2006.

12 Steve Fraser, *Every Man a Speculator*, 606.

13 Jeff Faux, *The Global Class War* (Wiley, 2006), 122.

14 Robert Rubin and Jacob Weisberg, *In an Uncertain World: Tough Choices from Wall Street to Washington* (New York: Random House, 2003), 11.

15 Robert Metz, "Wall Street Had a Hand in Mexico Bailout," *Tulsa World*, February 12, 1995

16 Rubin and Weisberg, *In an Uncertain World*, 11.

17 Michael Prowse, "The Rescuers," *The New Republic*, February 27, 1995.

18 Rubin and Weisberg, *In an Uncertain World*, 26.

19 Metz, "Wall Street Had a Hand in Mexico Bailout."

20 Prowse, "The Rescuers."

21   Louis Uchitelle, "Helping Hand Replaces Hands-Off Role in Asia," *New York Times*, January 1, 1998.

22   Bill Clinton, *My Life*, (Knopf, 2004), 642.

23   Paul Bluestein, *The Chastening: Inside the Cases that Rocked the Global Financial System and Humbled the IMF*, (Public Affairs, 2003).

24   Clinton, *My Life*, 643.

25   Jorge Castañeda, *The Mexican Shock: It's Meaning for the United States*, (New York: New Press, 1995), p. 202.

26   Uchitelle, "Helping Hand Replaces Hands-Off Role in Asia."

27   Ibid.

28   Nicholas Kristof, "Of World Markets, None an Island," *New York Times*, February 17, 1999.

29   Quoted in Faux, *The Global Class War*, 119.

30   Richard W. Stevenson, "Rubin Defends Asian Bailout and Asks Funds for IMF," *New York Times*, January 22, 1998.

31   Paul Bluestein, *The Chastening: Inside the Crisis that Rocked the Global Financial System and Humbled the IMF* (New York: Public Affairs, 2003), 185.

32   Bluestein, *The Chastening*, 9.

33   "U.S. Businesses Press For An Asian Bailout," New York Times, April 5, 1995.

34   Bluestein, *The Chastening*, 294–295.

35   Ibid., 325.

36   Paul Bluestein, *And the Money Kept Rolling in (And Out): Wall Street, The IMF, and the Bankrupting of Argentina* (Public Affairs, 2005), 158; Nicholas D. Kristof and Sheryl WuDunn, "Of World Markets, None an Island," *New York Times*, February 17, 1999.

37   Bluestein, *The Chastening*, op. cit.

38   Bluestein, *The Chastening*, 202; for information on stock prices, see Zhaohu Zhang, "The Impact of IMF Bailout on U.S. Bank Creditor's Equity Values: An Event Study of South Korea's Case," Doctor of Philosophy dissertation, Texas Tech University, May 2000.

39   Bluestein, *The Chastening*, 11.

40   Faux, *Global Class War*, 119.

41   Bluestein, *The Chastening*, 296.

42   "Financial Crises in Emerging Markets: Incentives and the IMF," Joint Economic Committee, United States Congress, August 1998.

43   Richard W. Stevenson, "Rubin Defends Asian Bailout and Asks Funds for IMF," *New York Times*, January 22, 1998.

44   Andrew Ross Sorkin, "Dimming the Aura of Goldman Sachs," *New York Times*, April 17, 2009.

45   Gary H. Stern and Ron J. Feldman, *Too Big to Fail: The Hazards of Bank Bailouts* (Washington: Brookings, 2004), 77.

46   Bluestein, *The Chastening*, 300

47   Nicholas Kristof and Sheryl Wodunn, "World's Market, None of them an Island."

48   Dave Kansas, *The Wall Street Journal Guide to the End of Wall Street As We Know It* (New York: Harper Business, 2009), 10.

49   Rubin and Weisberg, *In an Uncertain World*, 286.

50   Richard Bookstaber, *A Demon of Our Own Design: Markets, Hedge Funds, and the Perils of Financial Innovation* (New York: Wiley, 2009), 109.

51   Bluestein, *The Chastening*, 323.

52   Rubin and Weisberg, *In an Uncertain World*, 287.

53   Charles R. Morris, *The Two Trillion Dollar Meltdown* (Public Affairs, 2008), 53.

54   Mark Landler, "U.S. Hedge Fund Bailout Raises Asian Eyebrows," *New York Times*, September 29, 1998.

55   Stern and Feldman, *Too Big to Fail*, p. 17.

56   Louis Uchitelle, "Calculated Risk: U.S. and IMF Lead Push for Brazil Bailout Plan," *New York Times*, September 28, 1998.

57   Patricia Lamiell, "Latin America Ripe for Investments, Citigroup Unit Says," Associated Press archive, November 4, 1998; Paul Bluestein, "U.S., IMF Announce Plan to Avert Brazilian Crisis—Loan Package Totals $41.5 Billion," *Washington Post*, November 14, 1998.

58   Bluestein, *And the Money Kept Rolling in (And Out)*, 117–118.

59   Ibid., 146.

60   Quoted in Juliusz Jablecki and Mateusz Machaj, "The Regulated Meltdown," *Critical Review*, nos. 2–3 (Spring–Summer 2009).

61   Barry Eichengreen, "Bailing in the Private Sector: Burden Sharing in International Financial Crisis Management," *Fletcher Forum of World Affairs*, Winter–Spring 1999, 57.

CHAPTER 7: MINORITY MELTDOWN: A TALE OF TWO BUBBLES

1 Dave Kansas, *The Wall Street Journal Guide to the End of Wall Street As We Know It* (New York: Harper Business, 2009), 4.

2 Vikas Bajaj and Ron Nixon, "For Many Minorities, Signs of Trouble in Foreclosures," *New York Times*, February 22, 2006.

3 Anthony Ha, "Minorities Hit Hard by Foreclosure Crunch," *Hollister Free Lance*, May 3, 2007.

4 Edward Ericson, Jr. "Victim Mentality," *Baltimore City Paper*, July 30, 2008.

5 Kevin Kemper, "Foreclosures on Pace to Set New High in '05," *Columbus Business First*, September 19, 2005.

6 Kristopher S. Gerardi and Paul S. Willen, "Subprime Mortgages, Foreclosures, and Urban Neighborhoods," Federal Reserve Bank of Boston, Public Policy Discussion Papers, no. 08–6.

7 Rakesh Kochhar, Ana Gonzalez-Barrera, and Daniel Dockterman, "Through Boom and Bust: Minorities, Immigrants and Homeownership," Pew Hispanic Research Center, May 12, 2009, http://pewhispanic.org/reports/report.php?ReportID=109.

8 Michael Powell and Janet Roberts, "Minorities Hit Hardest as New York Foreclosures Rise," *New York Times*, May 16, 2009.

9 Dana Ford, "Minorities Hit Hardest by Housing Crisis," Reuters, November 26, 2007; Treisa Martin and Caitlin Watt, *Subprime Crisis: A Comprehensive Analysis from a Systems Thinking Perspective*, Kirwan Institute, Ohio State University, August 2008.

10 Todd J. Zywicki and Joseph D. Adamson, "The Law and Economics of Subprime Lending," *University of Colorado Law Review* 80 (2008).

11 Gerardi and Willen, "Subprime Mortgages."

12 Andrew Haughwout, Christopher Mayer, and Joseph Tracy, *Subprime Mortgage Pricing: The Impact of Race, Ethnicity, and Gender on the Cost of Borrowing* (New York: Federal Reserve Bank of New York, April 2009).

13 Stan Liebowitz, "New Evidence on the Foreclosure Crisis," *Wall Street Journal*, July 3, 2009.

14 Powell and Roberts, "Minorities Hit Hardest as New York Foreclosures Rise."

15 Quoted in William D. Cohan, *House of Cards: A Tale of Hubris and Wretched Excess on Wall Street*, (Doubleday, 2009).

16 Kansas, *The Wall Street Journal Guide to the End of Wall Street As We Know It*, p. XXX.

17 Alan Katz and Ian Katz, "Greenspan Slept As Off-Books Debt Escaped Scrutiny," Bloomberg News, Oct. 30, 2008.

18 Kansas, *The Wall Street Journal Guide to the End of Wall Street As We Know It*, 120.

19 Ibid.

20 Binayamin Appelbaum and Neil Irwin, "Congress's Afterhought, Wall Street's Trillion Dollars," *Washington Post*, May 30, 2009.

21 Quoted in Kevin Phillips, *Bad Money: Reckless Finance Failed Politics and the Global Crisis of American Capitalism,* (Penguin, 2009).

22 V. V. Chari, Lawrence Christiano, and Patrick J. Kehoe, "Facts and Myths about the Financial Crisis of 2008," Federal Reserve Bank of Minneapolis, Working Paper 666, October 2008; "Credit Crunch: What Credit Crunch? Au: Please verify title" Reuters, December 11, 2008.

CHAPTER 8: ROBIN HOOD COMES TO THE WHITE HOUSE: HOW BARACK OBAMA'S PLAN TO USE LIBERAL STATE CAPITALISM WILL CREATE THE NEXT BIG ECONOMIC MELTDOWN

1 Jeffrey Toobin, "Barney's Great Adventure," *The New Yorker*, January 12, 2009; "Fannie, Freddie Asked to Relax Condo Rules: Report," Reuters, June 22, 2009.

2 Claude R. Marx, "Rep. Johnson Introduces Measures to Include Credit Unions in CRA," *Credit Union Times*, March 18, 2009.

3 Quoted in Luke Mullins, "Barney Frank," *U.S. News and World Report*, March 1, 2009.

4 Michael Powell and Janet Roberts, "Minorities Hit Hardest as New York Foreclosures Rise," *New York Times*, May 16, 2009.

5 Alan Zibel, "Justice Dept. Says It Is Investigating Discrimination Against Minorities in Home Loans," Associated Press, July 25, 2007.

6 *Congressional Record*, (November 15, 2007), H13979, H13980.

7 Holman W. Jenkins, Jr., "Rainmaker Redux," *Wall Street Journal*, January 24, 2001.

8 Richard Muhammad, "Leaders Push Grassroots Agenda Forward," *Disclosure*, May–June 2005.

9   Tim McLaughlin, "Community Bank Finds Paranoid Smart Bet," *Boston Business Journal*, March 13, 2009; James Pethokoukis, "Yes, the Community Reinvestment Act Really Did Help Cause the Housing Crisis," *U.S. News and World Report*, June 25, 2009.

10 Mike Hall, "Obama Housing Plan 'Aims Straight at the Heartland,' " AFL-CIO Now Blog, Au: Need URL.

11 "Republicans Object to Stimulus Dollars for ACORN," FOXNews.com, January 27, 2009.

12 "Chicago-Based NTIC to Pay U.S. $550,000 to Settle Alleged Misuse of DOJ Funds to Lobby Congress," U.S. Department of Justice Press Release, June 3, 2009.

13 "White House Denies Charge by Attorney that Administration Threatened to Destroy Investment Firm's Reputation," ABCNews.com, May 2, 2009.

14 David Cook, "Obama: It's like AIG Has a Bomb Strapped to Them," *Christian Science Monitor*, March 18, 2009.

15 Stuart Varney, "Obama Wants to Control the Banks," *Wall Street Journal*, April 4, 2009.

16 Phillip Inman, "Goldman to make record bonus payout," *The Observer*, June 21, 2009.

17 Mark DeCambre, "Double-Dippers," *New York Post*, March 25, 2009.

18 "Not So Fast," *The Economist*, June 20, 2009.

19 "New Citi Chair: Bankers Aren't " 'Villains,' " CBSNEWS.com, April 7, 2009.

20 Thomas Cooley, "Moral Hazard on Steroids," *Forbes*, March 11, 2009.

21 Aaron K. Chatterji and Barak D. Richman, "Understanding the 'Corporate' in Corporate Social Responsibility," *Harvard Law and Policy Review*, Winter 2008.

22 Jonah Goldberg, *Liberal Facism*, (Doubleday, 2008).

23 Milton Friedman, *Capitalism and Freedom*, (University of Chicago Press, 2002)

24 Francis Cianfrocca, "Wealth Creation Under Attack," *Commentary*, June 2009.

25 Steven Malanga, "Obsessive Housing Disorder," *City Journal*, Spring 2009, 24–25.

26 Tom Borelli, "Congress, Corporate Lobbyists Creating Green Bubble," *Washington Examiner*, March 6, 2009.

27 Jonathan Fahey, "Free Juice!," *Forbes*, June 22, 2009.

28 Gabriel Alvarez, "Study of the Effects on Employment of Public Aid to Renewable Energy Sources," Universidad Rey Juan Carlos, March 2009; George Will, "Tilting at Green Windmills," Au: Appeared in what?, June 25, 2009.

29 Quoted in William Tucker, "The Next Subprime Mortgage Meltdown," *American Spectator*, February 17, 2009.

INDEX